THE NEW WORLD OF WORK

ENDORSEMENTS

Massive transformational change requires three fundamental ingredients: a clear direction, empowered leaders in effective roles, and a well-developed plan. Once the direction has been set, OD plays a critical role in ensuring that we deploy great talent in ways that allow a team to work together and really drive the change.

Leveraging organisational core competences requires the right leaders to be correctly deployed to drive performance. OD plays a critical role in allowing this to happen. Success is the only real change management technique that works, and leveraging what works is a great start for any large scale change.

The co-authors provide comprehensive reference points and some fresh OD thinking, with clear frameworks and practical checks.

Neil Maslen, Office Depot Europe, Chief Executive Officer and Chief Sales Officer
Contract Channel Europe

In today's complex corporate environment, simple solutions in real time are hard to come by. It always seems obvious until systems lock and structures and processes become static. This book on OD is a liberator of the thought process and a fit for purpose mediation in human and organisational optimisation.

Prof Frans Maloa, Associate Professor in the Department of Industrial and Organisational
Psychology, University of South Africa

Whether it is design or development is less important than ensuring that leaders are empowered to drive change quickly. Great leaders who are poorly deployed not only create frustration and place at risk the talent that is so key to change, but slow down the rate of change. In today's world we cannot afford for this to happen.

Peter Makapan, Head of Secretariat, Independent Remuneration Commission,
Presidency of the Republic of South Africa

"Agile is an attitude, not a technique with boundaries. An attitude has no boundaries, so we wouldn't ask 'can I use Agile here', but rather 'how would I act in the Agile way here?' or 'how Agile can we be, here?"
—*Alistair Cockburn*

First published in 2019.

ISBN: 978-1-86922-793-7

eISBN: 978-1-86922-794-4

Published by KR Publishing
P O Box 3954
Randburg
2125

Republic of South Africa

Tel: (011) 706-6009
Fax: (011) 706-1127
E-mail: orders@knowres.co.za
Website: www.kr.co.za

Typesetting, layout and design: Cia Joubert, cia@knowres.co.za
Cover design: Marlene de'Lorme, marlene@knowres.co.za
Editing & proofreading: Jennifer Renton, jenniferrenton@live.co.za
Project management: Cia Joubert, cia@knowres.co.za
Index created with TExtract/www.Texyz.com

THE NEW WORLD OF WORK

AN 'SOS' CALL TO MANAGEMENT

Edited by

MARK BUSSIN & CHRIS BLAIR

kr
publishing

2019

ACKNOWLEDGEMENTS

This book would not have been published without the extensive contributions of many very dedicated people.

Thank you to the contributors of the relevant chapters.

Many more of our colleagues, clients and students have inspired us and challenged our thinking – thank you to all.

We applaud the relentless quest for raising the bar of knowledge by all the 21st Century consultants that continually contribute to innovative discourse and the future way of work.

To Knowledge Resources, thank you for coordinating the production and marketing of this book.

A special thank you to Chris Blair for his insight.

A great thank you to Marina, Daniel, Kate, Genna and James for your inspiration and patience.

Dr Mark Bussin
Johannesburg, 2018
drbussin@mweb.co.za
+27 82 901 0055

FOREWORD

by Professor Margie Sutherland

In today's fast-paced world, cycle times are getting shorter and shorter. Our organisations are designed like massive military machines, all marching to the beat of one big drum. What is required, however, are pockets of excellence that are quick to respond and can move unhindered and with agility. Organisation design (OD) is a driver of organisational strategy, operational performance, employee commitment, job satisfaction and retention. It is thus a key element of performance management, and as such, needs to be deeply understood. It is simultaneously a highly complex field – half an art and half a science – needing deep understanding to inform the myriad decisions and trade-offs one has to make. The outcomes of these decisions have the potential to be either positive or negative for individuals, teams and organisations. This book will assist the student and managers to maximise the return on employee costs, which in many firms account for more than 50% of operating costs. In order to maximise the return on investment, the organisation structure must be defendable, efficient and effective. Organisation design intersects with many domains in organisational life: budgeting, cost control, annual reporting, human behaviour, process efficiency, marketing, market research, communication practices both internal and external, as well as productivity. It has its own lexicon, which needs to be mastered.

The current range of legislative, corporate and organisational reporting demands; the often-critical media reportage and exposés; stakeholder pressure for moral corporate governance; and demands for greater transparency increase the need for this book. Organisation design is a difficult skill that needs to be mastered as part of the career capital of human resource managers who are serious about their own futures and credibility. All executives who serve on, or aspire to serve on, boards of directors and remuneration committees, need a sound body of knowledge of organisation design practices to influence the success of the organisations they serve. This book provides the base for acquiring the knowledge, skills and worldview necessary for accountable leadership.

The insights in this book need to be put to good use and will provide a springboard for career and organisational success. Mark Bussin consistently contributes to the development of a host of business leaders and experts via corporate and consulting experience, wise counsel, writings, and hundreds of lectures, TV and radio interviews. He has upskilled a generation of HR professionals, helped define the field of practice, and made a significant contribution to the national level of excellence in the field.

As a young postgraduate student, Mark was given an article to read on "Super-leadership", in which he learned that to rise to great heights, one has to give away all that one knows to as many people as possible. This he has done tirelessly. He has informed the worldview of thousands of individuals and organisations. It is a great pleasure to see the fruits of his career made available to a wide audience in this well-written, usable and value-adding handbook.

Professor Margie Sutherland
Gordon Institute of Business Science, University of Pretoria

TABLE OF CONTENTS

List of Tables

List of Figures

ABOUT THE EDITORS

Mark Bussin is the Chairperson of 21st Century (Pty) Ltd, a specialist remuneration and HR consultancy. He has HR and remuneration experience across all industry sectors, and is viewed as a thought leader in this arena. Mark has held global executive positions for several multinational organisations, including mining, FMCG and financial services organisations. He serves on and advises numerous Boards and Audit and Remuneration Committees. Mark holds a Doctorate in Commerce. He has published or presented over 450 popular articles and papers and 45 peer reviewed journal articles. Mark is a guest lecturer at several universities and supervises Masters and Doctoral theses in the HR, Leadership and Reward areas. He is a past President of SARA (South African Reward Association) and a past Commissioner for the remuneration of Public Office Bearers in the Presidency. Mark tutors for WorldatWork globally. Mark enjoys flying Cessnas and loves his family time.

Mark can be contacted at: drbussin@mweb.co.za or on +27 (0)82 901 0055, or visit his website: www.drbussin.com.

Chris Blair, Chief Executive Officer of 21st Century (Pty) Ltd, has consulted to over 1000 organisations – both in Southern Africa and internationally. 21st Century is a leading management consultancy in Remuneration and Organisation Development. Chris holds a BSC Chem. Eng. (University of Witwatersrand), an MBA in Leadership & Sustainability (University of Cumbria) and is currently studying his PhD through the University of Lancaster. He is registered as a Chartered Human Resource (CHR) Practitioner with the South African Board for Personnel Practice (SABPP) and is also accredited as a Master Reward Specialist through the South African Reward Association (SARA).

Chris has served on the following Boards: Council member of Cape Peninsula University of Technology; member of the Remuneration Committee, Audit and Risk Committee, Joint Finance and Audit and Risk Committee, and Human Resources Committee of Cape Peninsula University of Technology; Chair of MEDO (Pty) Ltd; Board member of Huguenot College; Non-executive Director of NuQ (Pty) Ltd; Member of the Institute of Directors for Southern Africa Remuneration Forum; Trustee on a Sanlam Umbrella Provident Fund; Sub-committee member of the Investment Committee of Cape Peninsula University of Technology; member of the Accreditation Committee for the South African Reward Association; and as an advisor to numerous Remuneration Committees.

Chris can be contacted at: cblair@21century.co.za or on +27 (0)82 578 5932, or visit www.21century.co.za.

LIST OF CONTRIBUTORS

Belinda O'Regan (boregan@21century.co.za)

Bryden Morton (cblair@21century.co.za)

Dave van Eeden (dvaneeden@21century.co.za)

Lizette Bester (lbester@21century.co.za)

Morag Phillips (mphillips@21century.co.za)

OVERVIEW AND CONTEXT

The idea to write this book came from watching the news and real life events. Modern giant economies have military might with millions of troops, thousands of tanks, hundreds of stealth aircraft and plenty of giant warships. They are structured like big, well-oiled machines, which worked well for the last 500 years until along came small groups of insurgents, much the size of mosquitoes, relatively speaking, which besieged this giant machinery despite all its intelligence, weaponry and radar.

At the same time, I read a book called *Competing with Everyone from Everywhere for Everything*. Clients were also complaining that they just didn't feel Agile enough to compete. I joined the dots – our current organisations are structured like giant, well-oiled military machines with chain of command type management, but what is required is a much more Agile and nimble organisation that can reach any customer, anywhere and at any time.

The trigger to finally put this book together was when I was served a meal in a restaurant in Asia by a robot, having placed my order on an iPad on the table. The next event that really sealed it for me was checking my baggage in at Munich airport and there was not one human behind the almost 100 counters. I keyed my destination into the computer screen that faced me (not the humans that used to sit on the other side of the counter) and I hit a snag. My flight was Munich – Zurich – Johannesburg and I didn't know how to check my baggage from Munich all the way through to Johannesburg. I tried to find a human to assist me and had to walk quite a way to an information desk to ask them. It was a bit embarrassing because the answer was – the computer does it automatically – duhh...

This book is aimed at helping current and future HR practitioners and business leaders understand the complex subject of Organisation Design (OD) and the **HR implications**. We are aware that traditionally organisation design refers to organisation development, but this book is more about the design of organisations and the HR implications of the design. Our aim is to provide insight from lessons learnt on how the HR function needs to respond to a new organisation's designs and Agile structures.

The area of organisation design is increasingly important for business, HR and leaders as the VUCA world we live in faces dramatic demographic shifts, sustained market growth globalisation, and cultural and generational differences that are causing many organisations to re-think the way they engage with their people. HR policies, procedures and practices may not be the only part of the EVP today that helps to attract, retain and engage talent, but it is clearly important. Properly designed work and jobs can create value for all stakeholders. My son, James, is 6 years old and growing up with "screens", so his attention span and interests are huge red flags for organisations wishing to employ him. Work had better be interesting, challenging and exciting. We need to be ready with our employment practices!

The key features and layout of the book include:

- agility in the workplace, the rise of the contingent worker and attracting and retaining talent;
- organisation challenges and responses to agility;
- driving teamwork and rewarding teams;
- commensurate leadership and people practices;
- the key features and differences between robotics and artificial intelligence (AI); and
- workspace design, co-location and how the HR function should respond.

One of the main features of this book is the blend of academic rigour coupled with practical advice to illustrate the points we make. We include examples/cases where the authors or collaborators have had a direct hand in implementing the strategy or concepts explained in the book.

We trust you will find this handbook a useful tool to learn about the different aspects of organisation design and a reference guide for years to come.

CHAPTER 1

AGILE TAKES OVER THE WORKPLACE

INTRODUCTION

The 'New World of Work' is not a future concept, it is a current reality. Not only has the nature of work changed, but how we organise ourselves to optimally deliver the work is also evolving. For more than 100 years, organisations have been designed for scalable efficiency where functional teams have been constructed to run product design, engineering, manufacturing, sales, service, marketing, finance and human resources.[1] The mantra of 'standardisation' with the goals of 'cost and process-efficiency' have resulted in siloed structures and behaviours that inhibit customisation and customer centricity.

In this new world of work, globalisation and emerging technologies have reduced barriers to entry in many markets and the number of new entrants and non-traditional competitors is increasing. It is often easier for new entrants to be Agile and innovative as they are not reliant on legacy systems, existing structures, institutionalised ways of work and hierarchical structures occupied by people with long tenure and high levels of resistance to change.

The advent of 'Agile' is challenging traditional organisational structures and ways of working, which need to start competing with their smaller and more nimble competitors. The ability to act and think like a start-up, but back this with size, scale and an established brand, is powerful. There is no doubt that Agile has become the 'flavour of the day', and is fast becoming a value that most modern organisations subscribe to. The intention of this book is to explore this brave new Agile world of work and what it takes to shift mind-sets, behaviours, structures and practices to enable success.

WHAT IS AGILE?

Agile is defined as the ability to move quickly and easily, as well as having a resourceful and adaptable mind-set.[2] In the organisational context, Agile is a methodology/framework that was originally established for software development. Agile is not just for IT anymore, but is being used in HR, marketing and other disciplines and organisations in other industries. Most industry sectors have embraced Agile as a way of gaining a competitive advantage.

The History of Agile

In 2001, 17 people met in Utah, USA, to talk about software development and share ideas. This informal gathering turned out to be a seminal moment in the history of both IT and business in general. While discussing their mutual desire to find an alternative to "documentation driven, heavyweight software development processes", they penned the Agile Manifesto, changing the course of traditional development methodologies forever.[3]

Manifesto for Agile Software Development

We are uncovering better ways of developing software by doing it and helping others do it. Through this work we have come to value:

Individuals and interactions over processes and tools
Working software over comprehensive documentation
Customer collaboration over contract negotiation
Responding to change over following a plan

That is, while there is value in the items on the right, we value the items on the left more.[4]

The Agile Manifesto challenged traditional ways of software development and called for a different focus and prioritisation within the workplace. The Manifesto was accompanied by a set of Agile Principles, which further explain the paradigm shift.

Principles behind the Agile Manifesto

We follow these principles:

Our highest priority is to satisfy the customer through early and continuous delivery of valuable software.

Welcome changing requirements, even late in development.

Agile processes harness change for the customer's competitive advantage.

Deliver working software frequently, from a couple of weeks to a couple of months, with a preference to the shorter timescale.

Business people and developers must work together daily throughout the project.

Build projects around motivated individuals. Give them the environment and support they need, and trust them to get the job done.

The most efficient and effective method of conveying information to and within a development team is face-to-face conversation.

Working software is the primary measure of progress.

Agile processes promote sustainable development. The sponsors, developers, and users should be able to maintain a constant pace indefinitely.

Continuous attention to technical excellence and good design enhances agility.

Simplicity--the art of maximizing the amount of work not done--is essential.

The best architectures, requirements, and designs emerge from self-organizing teams.

At regular intervals, the team reflects on how to become more effective, then tunes and adjusts its behavior accordingly.[5]

How do Agile Methodologies work?

The main tenants of Agile methodologies are incremental delivery, team collaboration, adaptive planning, and continual learning.[6] Let's break these down:

Incremental delivery

Instead of building software and only delivering it at the end when it is 100% complete, an Agile methodology breaks the project down into smaller pieces of user functionality called user stories. It then prioritises these user stories and continuously delivers them in short two-week cycles called iterations.[7] The software is thus built incrementally, with each iteration eventually culminating in the overall deliverable/project goal.

Team collaboration

The base unit of production in an Agile methodology moves from the individual to the team (sometimes called the squad). It is arguably the power of collaboration within these teams that drives the success of the entire methodology. Teams are designed to be cross-functional to encourage communication between business owners and IT and leverage diversity of skill, knowledge and experience.

> **User Stories**
>
> A user story is the smallest unit of work in an agile framework. It is an end goal, not a feature, expressed from the software user's perspective.
>
> Each story consists of a few sentences in simple language that outline the desired outcome. They do not go into detail.
>
> User stories are critical as they keep the focus on the user and focus the team on solving problems for real users.
>
> Stories also drive creative solutions by encouraging the team to think critically and creatively about how to best solve for an end goal.[8]

It is important to differentiate between cooperation and collaboration in teams. Mahale explained the difference powerfully by using a simple mathematical analogy. When people *cooperate* they divide the responsibilities and identify touch points. In mathematical terms this is similar to 1 + 1 = 2. Essentially, each team member is doing what is expected of their role, but nothing beyond.

When people *collaborate* they build on each other's strengths and knowledge to create something that is exceptional and beyond their individual abilities. In mathematical terms this is similar to 1 + 1 > 2. Collaboration involves a lot of negotiating, challenging assumptions, and learning/building on each other's perspectives. The result of such collaboration is that barriers and departmental silos are naturally overcome, enabling the team to be greater than the sum of its parts.[9]

Although team members still have core competencies, roles blur on Agile projects. Everyone pitches in and does whatever is required regardless of their role or title. One of the greatest benefits of an Agile approach is the removal of hierarchy and the flattening of organisational structures. The role of leadership is to remove obstacles and facilitate the work, as opposed to directing and controlling the work. The fundamental belief is that teams should be self-managing and that the people "in the work" are best positioned to come up with solutions "for the work". The reliance on leaders and management consultants to come-up with all the answers is laid to rest, albeit not always peacefully!

Collaboration does not always come up naturally and teams often need coaching to encourage a growth mind-set. A growth mind-set allows people to thrive when faced with challenges by promoting the belief that their basic qualities and competencies can be cultivated and improved through their efforts. One of the biggest challenges people face when moving from traditional ways of work to an Agile methodology is moving from a narrowly defined role involving a skill set honed over many years to a self-organising, cross-functional team.[10] In order to adapt to the new ways of working espoused by Agile methodologies, people need to be adaptable and see the requirement to learn new skills and develop new competencies as a positive growth opportunity.

While coaching and experience can improve collaboration, there are also more concrete steps that can be taken to foster a collaborative work environment. Co-location of team members plays a significant but not essential role in driving collaborative practices. Technical tools and collaboration platforms can play a key role in bridging the gap when team members cannot be co-located and, also allow for the creation, sharing and updating of documentation in real time by multiple team members. The rituals and ceremonies involved in Agile methodologies are also designed to promote close collaboration, for example, teams usually have a daily stand-up or short meeting to touch base and make sure everyone is aligned.

Adaptive planning

Traditionally change has been frowned upon in software projects, especially when it arises late in the development process, yet the use of incremental delivery allows for change to be adaptive and for the cost of change to be managed. The measures of success are far more customer-centric in an Agile methodology; a project may come in on time and on budget, but if it misses the mark because customer requirements have changed, the results are of no or decreased value to the customer. Adaptive planning allows teams to embrace change and pivot as and when required to make sure customer requirements remain the key measure of success.

Those who do not understand Agile methodologies well and are used to long-term planning often argue that there is a lack of planning in Agile. This is simply not true. Agile ceremonies include:

- daily planning with daily stand-ups;
- bi-weekly planning with iteration/sprint planning meetings; and
- release planning, where teams decide what to release every three to four months.[12]

There is a great deal of planning in Agile, but it is continuous adaptive planning. Plans are expected to change and this is not seen negatively.

Continual learning

> **The three simple truths of Agile**
>
> 1. It is impossible to gather all the requirements at the beginning of a project. Understand that requirements are meant to be discovered and that not proceeding until all are gathered would mean never starting.
>
> 2. Whatever requirements you do gather are guaranteed to change. Accept and embrace change – adapt when necessary and move on.
>
> 3. There will always be more to do than time and money will allow. Having a to-do list that exceeds your time and resources to deliver is the normal state for any interesting project.[11]

Working in cross-functional teams with reduced administrative work, meetings and documentation leads to increased speed-to-market and an exponential increase in the pace of delivery. To support this pace, a learning culture is required. People working in Agile ways will need to have easily accessible learning material and will, more importantly, need to develop a passion and desire to learn continuously. In the past we placed value on deep expertise in mono-lines or content domains. Today we simply do not have the luxury of having teams of deep experts all working on the same product/feature. The ideal Agile team comprises 8-12 people, and time and cost constraints preclude large teams of specialists being brought in to solve for ongoing development. Despite the resource constraints, where there is a lack of multi-skilled people, teams become too big to collaborate effectively.

The call for multi-skilled Agile practitioners has resulted in a need to create and curate on-demand learning that serves both general Agile development as well as role-based deep learning needs. The learning journeys should include a variety of learning methods and experiences such as bite-sized nuggets/nano-learning, on-line tutorials, communities of practice, workshops, seminars, videos, games and simulations, coaching, mentoring, action-learning, job-rotation, and internships.

One of the most important roles of a leader in an Agile environment is to create and promote a culture of learning and trust. A learning culture was described by Gill as follows:

> "A 'learning culture' is a community of workers continuously and collectively seeking performance improvement through new knowledge, new skills, and new applications of knowledge and skills to achieve the goals of the organisation. A learning culture is a culture of inquiry; an environment in which employees feel safe asking tough questions about the purpose and quality of what they are doing for customers, themselves, and other stakeholders."[13]

Admitting that you do not have the answer to a problem or that you have failed is much more productive that continuing on a path where psychological safety prevents you from asking for the help or development required. Enabling a learning culture sounds a lot easier than it is in reality.

Continuous learning can be considered a destabilising force to existing culture. Questioning long-held assumptions can be risky in an organisation that values stability over learning.[14] In addition, while many organisations say that failing fast is encouraged, the impact on performance management, rewards and career progression demonstrates otherwise.

CASE STUDY

Crowd-sourcing technical training to build a continuous learning culture

General Dynamics C4 Systems develops and integrates communication products that deliver vital information for military, homeland security, and public safety professionals in the United States. Jeff Plummer describes an intervention in an area where high performing teams build collaborative battlefield management systems:

In an attempt to build a continuous learning culture in an environment where time, cost and challenging the status quo were limiting factors, a crowd-sourcing technical training model was developed. In a nutshell, engineers were asked to teach others the concepts that they themselves found interesting.

Interested in Teaching	Interested in Learning	Upcoming Training
Intro to Similarity Classifiers	**Hadoop**	**Apache Flex Deep Dive**
How does Amazon know what books you might like? How does Pandora know what songs you want to hear? Find out!	Linearly scalable platform for analysing large data sets/	5 2-hour sessions covering · Language basics · RIA constructs including charts, tables, and more · Mobile development with Adobe AIR
Interested?	Chad Tyler *Interested?* Nat Wetzel Lisa Hunter	*Interested?*
Jonathan Nancy Keith Terry Kyle Searcy Amber Christensen	Jim Macrepol Rob Umberger Terry Ford	John Smith Jane Doe Martin Towler Bob Robertson

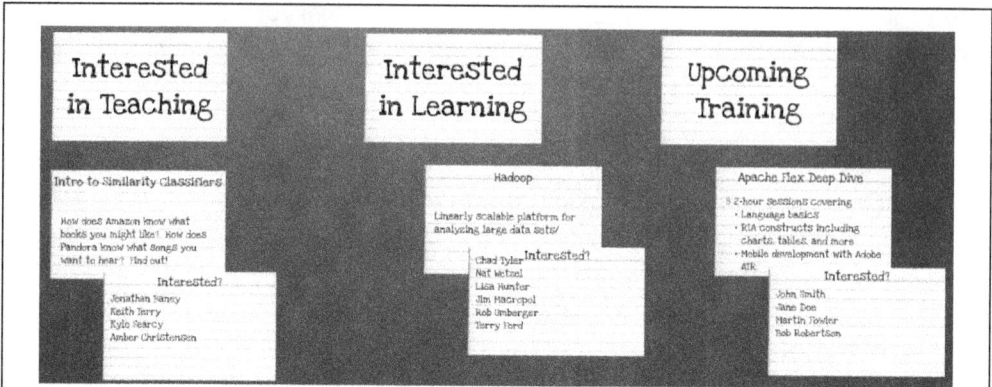

The net result was when enough people signed up as being interested, someone would find the motivation to put together the training. A person might not be willing to put together 8 hours of native Android training ordinarily, but when that person sees a large number of their peers or people they respect wanting to learn, it proved to be a powerful motivator. General Systems wanted a learning format that did not add to people's current workload, and made every second of the training feel valuable. They settled on the "brown bag lunch" for topic introductions, and "deep-dives" for more in-depth training:

Brown Bag Lunch: everyone brings a brown bag lunch to a conference room to attend a 1-hour presentation on a particular subject. The format is long enough that someone can give a good introduction to a topic, and short enough that people don't feel like they had to sacrifice anything to attend.

Deep Dives: the deep dive is a training session that takes two hours, one day a week, for several weeks. The idea is people will go to a lab in two hour increments, and gain a deep understanding of a technology by doing interactive exercises. Two hours per week is not enough to negatively impact people's normal workload, but it is long enough to actually walk through one or more small exercises. Over the course of several weeks, it is possible to train people to become proficient with technologies like HTML, Native Android, or Apache Flex.

Training champions were identified to help keep the initiative alive. These champions were the people who bought into the concept completely and were willing to throw together a brown bag lunch seminar at the last minute just to keep the momentum moving. The champions were also the people who were out encouraging their peers to give talks and pushing others to keep the training initiative moving forward.

The results of the training initiative were impressive and immediate.

Employee happiness: employee overall happiness rose significantly with the training initiative. Their fear of the unknown disappeared because they realised these new technologies are not hard to learn, and much of their other technical skills directly apply.

Attrition was reduced: one of the primary reasons intelligent people leave their jobs is because they aren't growing professionally. With an environment where people are constantly learning new concepts and technologies, that reason is significantly diminished.

New innovation: one of the welcome side effects of a well trained staff is an influx of new ideas.[15]

Agile Methodology Myths

As with any new methodology that tries to change culture and threatens the status quo, a number of myths about Agile have arisen. The myths generally arise from a lack of understanding of how Agile methodologies are practiced on the ground. Some of these myths are described and refuted below.

Agile is a Silver Bullet

There is nothing magical about Agile and you can fail just as badly using Agile as you can using any other methodology. Agile brings your development team and customer closer together and gives your people permission to do their best work and be accountable for the results. There are no guarantees of success.

Agile is Undisciplined

In truth, Agile is a very disciplined way of delivering software:

- You have to test.
- You have to get feedback.
- You have to regularly release software.
- You have to change and update the plan.
- You have to deliver bad news early.

And of all this happens at pace with incremental results being visible and transparent throughout.

Agile is Anti-Architecture

Agile has responded to big, complex, expensive, hard-to-maintain systems by promoting an attitude of simplicity. Agile is not anti-architecture, but it does push back on adding complexity where it is not needed.

Agile is Anti-Documentation

Agile is not anti-documentation but it does not encourage documentation for documentation's sake. Documentation is treated like any other deliverable on an Agile project. It gets estimated, sized, and prioritised like any other user story. Face-to-face communication is preferred to the written word when dealing with your people and your customers.

Agile Requires a lot of Rework

There is indeed rework in Agile but this comes in two forms:

- There is rework as a result of changing or refined customer requirements. This is a good thing as you want your product to be as close to real customer requirements as possible.
- There is rework of the software development teams as they discover better ways to design the software.

While both of these forms of rework are inherently good, they still need to be managed and balanced. You cannot keep changing your mind or design – at some point you have to release the product/feature. Agile deals with this tension by empowering both sides with the power to iterate, so long as they work within the project's means.

Agile does not Scale

Scaling is difficult – there is no way to easily coordinate, communicate, and keep large groups of people all moving in the same direction towards the same cause. To scale Agile, you must enable autonomy for the team, while ensuring alignment with the organisation. The critical building blocks are clearly defined ownership and a culture of trust. Once you have this foundation in place, you will find that Agile can scale very well.

Agile does not require a Roadmap

Organisations and teams following an Agile approach absolutely know where they are going... and the results they want to achieve. Recognising change as a part of the process (an agile approach) is different from pivoting in a new direction every week, sprint, or month.

Adapted from Rasmusson[16]

APPLYING AGILE PRACTICES TO THE GENERAL WORKPLACE

At the core of the Agile Manifesto is the notion of customer-centricity or putting the customer first. Many, if not most, traditional organisations are internally focussed. They work in 1-3 year budget cycles; require detailed planning of the entire project before any business case can be approved; and value extensive documentation and project management practices to ensure a solid audit trail. As a result, they are slow and inefficient. While there is much talk of the "customer being king", the reality is that when traditional structures and methodologies are employed, it is difficult to be responsive to rapidly evolving customer needs. This is exacerbated in highly regulated industries like banking and finance.

Process-driven business practices were developed in a time when change was moderate and managed through careful, deliberate and long-term planning. As the new world of work has become a reality, exponential change has challenged the efficacy of process-driven practices. Customers demand instant gratification, and with increasing and easily accessible competition, if your product or service cannot keep up, there are many alternatives. Customers are no longer as wary of switching products and services, and in many cases, 'new' is considered exciting and appealing. Even with brands that depend on brand loyalty, there is still an expectation that to keep customers interested they need to innovate, or at least copy, as fast as the competition.

In the new world of work, organisations that do not keep up are left behind, or even worse, cease to exist. More than 89% of the companies on the original Fortune 500 List from 1955 have either gone bankrupt, merged with (or were acquired by) another firm, or they still exist but have fallen from the top Fortune 500 companies list for one year or more. This demonstrates that there has been a lot of market disruption, churning, and creative destruction over the last six decades. This is great news for the consumer who is the ultimate beneficiary of the dynamism and innovation that characterises a consumer-oriented market economy[17], however it is not great news for organisations that are fighting to remain relevant, profitable and innovative.

The most commonly believed impetus for the increasingly high turnover on the Fortune 500 List is rapid technological development and the challenges that go with it.[18] Intense competition is now the order of the day. Organisations are understandably concerned about their sustainability and longevity and are seeking new ways to deliver value to their customers, while containing costs and fighting the never-ending war for talent. Agile methodologies represent a potential new way of work, for the new world of work.

The adoption of Agile has thus gone beyond software development and is now a driving force within many organisations across industries, sectors and geographies.

Agile methodologies promise to establish a culture and environment where solutions emerge. The focus is on "delivering good products to customers by operating in an environment that does more than talk about 'people as our most important asset' but actually 'acts' as if people were the most important".[19] Many argue that Agile is far more than just the adoption of a particular framework or process. Agile is seen as a mind-set guided by a set of principles. It is about how you collaborate with your customers, stakeholders and fellow team members and how you think about your work and approach.

Ahmed Sidky[20] illustrated the way in which Agile should be interpreted, clearly delineating the difference between Agile as a mind-set and Agile as practice – see Figure 1.1 below.

AGILE IS A MINDSET	DESCRIBED BY 4 VALUES	DEFINED BY 12 PRINCIPLES	MANIFESTED THROUGH UNLIMITED NUMBER OF PRACTICES

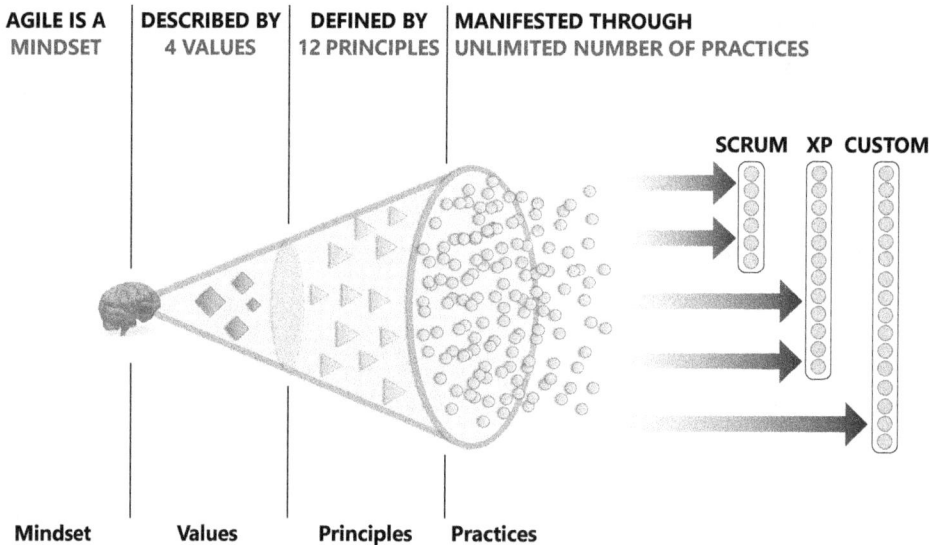

SCRUM XP CUSTOM

Mindset   **Values**   **Principles**   **Practices**

Figure 1.1: Ahmed Sidky's Agile mind-set[21]

There is no reason why an Agile mind-set cannot be applied to any organisation; the overall change required is fundamentally a culture change. Organisations that successfully adopt an Agile mind-set learn to trust and respect their people and focus on producing customer-centred products, whilst working in an inclusive environment where there is a pervasive sense of psychological safety. People can ask for help when they need it; they can fail-safe and try again to seek improvements. It is about expecting and embracing change and adapting to the changing needs of your customers.

IMPLICATIONS FOR DEVELOPING COUNTRIES

McKinsey forecasts that by 2025, nearly half of the Fortune Global 500 companies will come from emerging markets.[22] There is no doubt that emerging markets are changing where and how the world does business. Increasingly skilled labour and the emergence of a huge and prosperous middle class are contributing to predictions that future economic growth in developing countries will surpass that of developed countries.

Agile methodologies foster and enable innovation, and in developing countries, there is an opportunity to use innovation to leapfrog into the best and most advanced technologies.[23] There is undoubtedly a role for both Agile methodologies and an Agile mind-set in developing countries, but cognisance must be taken of the people and technology challenges. A major challenge from the people perspective is sourcing and retaining the right talent to work in cross-functional teams. Many of the roles in Agile are new (such as Agile Coaches and Scrum Masters), thus there is a need to up-skill and develop proficiency before there will be enough people with the requisite knowledge and capability to do the work. In addition, many of the technical roles (such as engineering leads and multi-stack developers) are considered both scarce and critical skills globally. Competition for these resources is fierce and there is a growing shortage of skills as more organisations decide to go Agile. There is thus a great risk in making a decision to adopt or scale Agile without conducting detailed Strategic Workforce Planning and making sure you have access to the right talent.

From an environmental/technical landscape perspective, solutions are not easily deployable across developing countries without significant business model adaptations to cater for particularly high social and economic constructs. There is a significant amount of customisation required to adapt to local conditions, from infrastructure challenges to affordability and education.[24]

The use of Agile methodologies is thus not an easy or simple solution for developing countries, but there is evidence that where an Agile mind-set and Agile practices are being implemented, there are signs of success. It is important to note that it is most often not an entire organisation that adopts Agile, but rather the change/delivery arm of an organisation. This usually involves IT, design and certain pockets of business. Agile practices thus coexist with traditional/waterfall practices and it is frequently the interplay between these old and new ways of work that cause friction. For example, you cannot fund Agile product development from traditional annual budgets; a funding pool has to be made available for quarterly business reviews.

This takes us back the assertion that Agile is more about culture than it is about process. If an organisation has the appetite to challenge existing practices, invest in its people and build a culture of trust, Agile can be successfully adopted regardless of whether the organisation is in a developing or developed country.

CASE STUDY

Successfully scaling Agile in developing countries (Standard Bank South Africa)

Standard Bank is the largest African banking group and is based in South Africa. For more than 152 years, the bank has served the continent and is now present in more than 20 sub-Saharan countries. Standard Bank was looking to improve service quality, efficiency, and employee morale, but previous efforts to scale Lean-Agile beyond a few teams had stalled.

To support its goals, the bank turned to the Scaled Agile Framework® (SAFe®) and gained backing from executives to move ahead with deploying it. Prior to rolling out SAFe, Standard Bank initiated various culture initiatives to start driving the change in behaviour of leaders and teams, and launched proofs of concept.

The bank took a number of steps to stretch out of its comfort zone:

- They pulled cross-functional teams together and began delivering on a cadence.
- The Internet Banking and ATM teams modelled breaking work down into smaller, more manageable pieces and demonstrated to stakeholders the work completed during the sprint.
- Business and IT stakeholders joined these showcases to provide feedback to the teams.
- They switched their work attire from suits and ties to jeans.
- They began running off-site sessions with IT to define culture themes, change guilds, and more.
- They initiated DevOps initiatives prior to the SAFe implementation which were formalised during the roll-out.

Additionally, the bank set a clear vision for the future of the organisation. At the top, leaders aligned around a common understanding of goals and key performance indicators (KPIs), and emulated Silicon Valley tech leaders on the kind of change and coaching culture required.

At the lower levels, the development community participated in defining the future state of the bank. Standard Bank also empowered employees to design their own culture as a group to achieve true ownership.

Prior to launching the first Agile Release Train (ART), Standard Bank portfolios embarked on an outside-in model, moving away from the traditional project structures into a SAFe design construct forming cross-functional teams, programmes, and portfolios. The bank set a milestone for the first of July 2016 for teams to co-locate, work from a backlog, and establish visual management of work and self-regulated teams.

With the outside-in design taking shape, Human Capital with support from the Group CIO started a programme that focused on re-skilling individuals to repurpose them as software engineers, quality engineers, or user experience analysts. Once they passed the aptitude test and went through the programme, they were placed in a feature team. As a result, the organisation now has more people getting the work done versus managing it.

From 1st July 2016 through February 2017, Standard Bank trained approximately 1,200 people on Leading SAFe in preparation for its first Programme Increment (PI) planning meeting in January 2017. These days, with more than 2,000 people trained on Leading SAFe, Lean-Agile practices and SAFe are key parts of Standard Bank's strategic plan. The move to SAFe delivered a number of benefits, both qualitative and quantitative. Standard Bank succeeded in breaking down silos and improving dependency management. They removed complexity and reduced cost, while building more. Business people now prioritise work and budgets to account for IT change.

The bank notes significant gains within some of the more mature teams or portfolios:
- Time-to-market reduced from 700 to 30 days.
- Deployments increased from once or twice a year to monthly.
- Productivity increased 50%.
- Costs decreased by 77%.
- Predictability is now at 68%.
- Organisational health improved by 12 percentage points from 2013 – 2016.[25]

CONCLUSION

As organisations consider their own existing or potential Agile transformation journeys, it is clear that there is no one-size-fits-all approach to adopting or implementing an Agile approach. Every organisation has different needs, constraints, and requirements. Simply applying Agile ceremonies and practices will not lead to success – an Agile mind-set and culture change is required. We need to find new ways of work for the new world of work and Agile offers a tried and tested solution that drives customer-centricity and speed-to-market. The good news is that thanks to the exponential acceleration of technology coupled with demographic and other economic shifts, there are many new ways to get work done. The bad news is that change is hard, especially culture change. As Agile continues to take over the workplace, we will hopefully see more organisations committing to an Agile mind-set and taking on the culture change challenge with enthusiasm and tenacity.

CHAPTER 2

THE RISE OF THE CONTINGENT WORKFORCE

Belinda O'Regan

INTRODUCTION

The ever-changing economic, environmental, social and political landscape is constantly challenging organisations to adapt and respond. The focus on the agility of organisations is highlighted by the growth in the start-up and accelerator models that are prevalent globally. Organisations as we know them, although often promoting the need for agility, are most often still structured in a manner that suits post industrial revolution foundations and concepts. The mind-set of how we structure organisations is slow moving and affected or limited by counteractive factors such as labour laws that impact the move to individuals being 'contingents' with very few employee protections. The argument, however, may be that we are not seeing the rise of the contingent workforce that has already begun, but rather the adoption and entrenchment thereof. The world of work is ever changing, with one of the most apparent and obvious trends being that of a more fluid and flexible workforce.

As Wilhelm Crous wrote: "Agile methodologies are transforming management and organisations."[26] They involve new values, principles, practices and processes, and are a radical alternative to command-and-control style management. This is a key foundation for why and how organisations are finding themselves unable to compete in a market if they are not fluid and Agile. The contingent workforce allows an organisation to expand and contract during times of opportunities and downturns.

DEFINING THE CONTINGENT WORKFORCE

A contingent workforce may be defined as "a labour pool whose members are hired by an organization on an on-demand basis. A contingent workforce consists of freelancers, independent contractors and consultants who are not on the company's payroll because they are not full-time employees of the organization. Organizations can hire a contingent worker directly or from a staffing agency. Such workers are usually added on an ad-hoc basis to a company's workforce and work either onsite or remotely. They generally receive fewer (if any) benefits and less pay than full-time workers, according to the U.S. Department of Labor, and are less likely to be protected by labor and employment laws".[27]

For the individual contributor, i.e. the "employee", the shift from having a job for life to shorter job stints to freelance work may have arisen from the uncertainty the economy has left them with, as well as the idea that having one job or career for life may no longer be in line with their values and motives for why they choose to work.

In this chapter we look at the contingent workforce as the group of people who are available for short term projects; they support organisational scalability and generally provide independent services to an organisation without becoming part of the payroll or structure, except on a pay for use basis.

THE DIFFERENCE BETWEEN THE TRADITIONAL WORKFORCE AND THE CONTINGENT WORKFORCE

Table 2.1: The differences between the traditional workforce and the contingent workforce

Traditional/Standard Arrangement	Contingent/Non-Standard Arrangement
Defined long term contract.	Contracts are generally short term and less defined or robust. Generally "zero" hour contracts but the understanding is that should a requirement arise, they would be contracted.
Anticipated continuous relationship.	Relationship is project-based; it may continue but is not guaranteed or expected.
Defined job requirements that are usually limited to one function/role.	May have a number of differing 'job titles' depending on offering.
Set and capped remuneration.	Remuneration is based on projects and negotiated.

Traditional/Standard Arrangement	Contingent/Non-Standard Arrangement
Set hours (although companies expect more than set hours).	Hours are set but more project-based; some contingent workers have zero hour contracts but are on call for defined projects.
Benefits and rewards.	No benefits offered by the company generally, but contingent employees find benefits in flexibility, variety of work, uncapped remuneration etc.
Working space and resources such as a PC provided by the company.	Have own infrastructure in most cases.
A manager – in charge of the function and the type of work assigned. Manages the quality of outputs as well as focusses on the wellbeing and motivation of the employee including development and career growth. A team – a group of people with defined functions to complete an assignment, project or performance in a day-to-day function.	A client – has a requirement and requests a proposal from the freelancer in order to complete the work. Will sign off when happy but generally does not manage the provider.
A team.	A project team/independent workers.
Expectation of the company promoting their brand and careers (employer-driven).	Promotion of own brand and career (person-driven).
Are paid via traditional payrolls.	May submit invoices for work done and manage own tax payments.

THE HISTORY AND EVOLUTION OF THE CONTINGENT WORKFORCE

The rise of the contingent workforce is likely to have been driven as much by the participants in the workforce as the agility requirements of businesses. Millennials expect a different working experience to what their parents and grandparents accepted; not only do they require a differing environment, but they are often unlikely to associate themselves with just one brand, especially if that brand is not congruent with their values and principles.

A key driver of the contingent model is that of flexibility not only for the employer, but the "contingent" themselves. Contingent employees are interested in flexibility not only of the working environment and set up, but also of the work they are exposed to and involved in.

The contingent model appears to have benefits for both parties, with the employer being able to react to large scale projects and opportunities without having the burden of a defined and expensive organisational structure. In a study by OCG it was found that: "One of the biggest impacts of the contingent workforce is the creation of flexibility for both employee and employer. For employees, working on a temporary or contract basis ensures that they have full control of their career, and allows them to do a wide variety of work, often for a range of clients. For employers, having a flexible workforce means that an organisation can easily scale their labour capability to the level they require. If there is a high demand for work, they can easily add more people with the necessary skills. On the other hand, if they are experiencing a work shortage, they are able to easily reduce costs as contingent workers are not as difficult to release as permanent ones. The access to this large contingent talent pool makes the management of large projects far easier, and less costly. Our survey indicated that this was the main reason for engaging contingent staff, with 17.8% of respondents indicating that this was the main benefit of contingent workers to their company".[28]

A study conducted by independent research firm Edelman Intelligence and commissioned in partnership by Upwork and Freelancers Union predicted in 2017 that by 2027 the majority of the US workforce would be contingent. In *Freelancing in America: 2017* it was also found that freelancers believe their career is more stable than a day job as they have a diversified portfolio of clients rather than a job with one employer. The concept of an employer for life that will look after you and your family is long in the rear-view mirror. It is about our own destiny now, hence the move to a contingent model.[29]

Dooley provided a concise and informative overview of the drivers and evolution of the rise of the contingent workforce, which is illustrated below.[30]

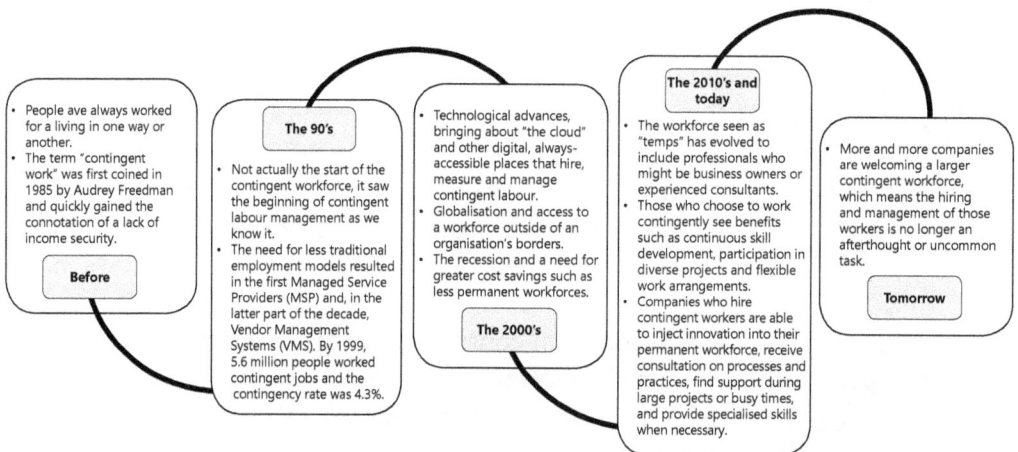

Figure 2.1: The evolution of the contingent workforce

THE DISRUPTERS IN THE WORLD OF THE CONTINGENT WORKFORCE

"Uber, the world's largest taxi company, owns no vehicles. Facebook, the world's most popular media owner, creates no content. Alibaba, the most valuable retailer, has no inventory. And Airbnb, the world's largest accommodation provider, owns no real estate. Something interesting is happening", noted Tom Goodwin, Senior Vice President of Strategy and Innovation at Havas Media.[31]

A key disrupter of the traditional sourcing and talent acquisition models are online employment platforms. These take several forms, from the providers of platforms where employers can search for talent and often hire in a more traditional way, to the true online platforms that assist organisations in the resourcing of their contingent models. These platforms may be seen as the "Uber" of the world of employment. They offer a massive workforce for use on demand without actually employing such individuals; they rather provide the platform to match up employers with transient/contingent workers.

Online talent platforms are increasingly connecting people to the right work opportunities. By 2025 they could add $2.7 trillion to the global GDP, and begin to ameliorate many of the persistent problems in the world's labour markets.[32]

EXAMPLES OF KEY PLAYERS IN THE INDUSTRY

Upwork, formerly Elance-oDesk, is a global freelancing platform where businesses and independent professionals connect and collaborate remotely.[33] Upwork has 12 million registered freelancers and 5 million registered clients. "Three million jobs are posted annually, worth a total of $1 billion USD".[34]

Freelancer is a crowd sourcing marketplace that connects over 31 million employers and freelancers globally from over 247 countries, regions, and territories. Freelancer allows potential employers to post jobs that freelancers can then bid to complete.[35] "Founded in 2009, its headquarters is located in Sydney, Australia, though it also has offices in Southern California, Vancouver, London, Buenos Aires, Manila, and Jakarta".[36]

"**Fiverr** is an online marketplace for freelance services. Founded in 2010, the company is based in Tel Aviv and provides a platform for freelancers to offer services to customers worldwide... As of 2012, over three million services were listed on Fiverr."[37]

The **Skyword** platform puts content at the core of all marketing activities. Skyword360 is designed to ensure that all cross-channel planning, content creation, and activation aligns with an enterprise's overall content strategy and marketing goals. "With an international freelance community from over 46 countries and the ability to translate the platform into 14 languages, content operations can be seamlessly executed at scale."[38]

EngineerBabu is a boutique of freelance tech experts who take responsibility from the inception to the completion of a project. The team is comprised of 45 in-house members and 20 exclusive partners, who work on technologies such as Angular2, AWS, and Magento.[39]

PeoplePerHour is a marketplace that connects small businesses and freelancers from all over the world in a trusted environment where they buy and sell services to each other.[40]

MOST POPULAR CONTINGENT SKILLS

Not all job types lend themselves to a contingent model. For example, the service industry requires a set and defined model of employment to ensure continuation of services. Although contractors and temporary staff may be employed, that is not the model we are focusing on in this chapter.

The following are the top 25 in-demand freelancer skills:[41]

Table 2.3: Top in-demand freelancer skills

1. Natural language processing	9. Brand strategy	18. Bluetooth specialist
2. Swift development	10. Business consulting	19. Stripe specialist
3. Social media management	11. Machine learning	20. SEO/Content Writing
4. Amazon Marketplace Web Services (MWS)	12. 3D rendering	21. Virtual assistant
5. AngularJS development	13. Zendesk customer support	22. Immigration law
6. MySQL programming	14. Information security	23. Accounting (CPA)
7. Instagram marketing	15. R development	24. Photography/video editing
8. Twilio API development	16. User experience design	25. Voiceover artists
	17. Node.js development	

IS THERE A NEW PSYCHOLOGICAL CONTRACT?

In traditional employment relationships there is what is termed a psychological contract. This contract was defined by Reber[42] as "the unwritten set of expectations that exists between the persons in a relationship, the members of a group, the people who work for an organisation etc. The term is most often used in industrial/organisational psychology, where it includes the levels of performance that each member of an organisation is expected to reach and each member's own expectations with respect to salary, advancement, benefits, prerequisites etc. Moreover, such nebulous components like the quality of life, job satisfaction, personal fulfilment and the like are implicitly part of the contract".

In the contingent world the psychological contract is turned on its head. The individuals themselves are now accountable for the diversification of opportunities in order to support their own expectations and growth needs. The employer is no longer tacitly responsible for these, although they possibly expect an even higher level of performance as they are not bound to retain services or engage on projects again should performance not be up to par. The concept of an employee lacking motivation to complete their work and being provided with support and mentorship from the organisation is not relevant in the model of contingent work. With this concept in mind, the thought may be that the contingent employee could feel isolated and disengaged as they have no support, foundation or sense of belonging. However, do they want to 'belong'? Where do they find the motivators and drivers for engagement? Are they engaged with the organisation or the work?

The answers to these questions may come from people's need for freedom. Even though they may be foregoing the 'security' of a predictable pay-check, they are gaining the freedom to work in different fields, on new projects and to define their own models of work. The new psychological contract may possibly be defined as a relationship that is transient in nature, whereby individuals come together for the successful completion of a specific task/project and then part ways with no commitment to each other. Each participant has defined what they need to get out of the relationship from an expectation perspective relating to financial rewards, which are completely defined and not movable unless the scope changes or rewards are not financial in nature but altruistic, personally satisfying or challenging. The intangible elements of the relationship to ensure performance is the possible consequence of being poorly rated on work platforms where these relationship matches are made, and hence the possible impact of not acquiring future transient projects and a tarnish on one's own personal branding. Engagement is with the work or project and not the organisation.

ADAPTING HR PRACTICES TO ACCOMMODATE THE CONTINGENT WORKFORCE

The traditional HR processes and policies are not necessarily applicable to the new workforce, yet most organisations are not effectively planning their contingent workforce requirements and thus are not building models to accommodate new workers and create a flexible yet less transient relationship. This type of work arrangement has its benefits in that the organisation does not have sunk costs, but they also do not have an ongoing contract with people who have now built their capabilities and competencies. David Brown, a partner at Deloitte Consulting, noted that it is important to create some form of connection with the contingent employee to keep them connected with the organisation so that they may return for further projects.[43]

Dylan White, a Senior Partner at Denovo, advised that HR needs to build in clear policies and processes for managing the workforce, with a mind-shift change regarding the concept of revenue generation with each project and an even stronger understanding of the business objectives and scope of the project. He added that there is still a need to include the contingent workforce in induction and training processes about the organisation to instil company values and processes.[44]

An additional factor that organisations traditionally have not managed effectively is tracking contingent worker costs as well as monitoring productivity. The OCG study –*The Rise of the Contingent Workforce* – surveyed a number of organisations and found that 75% of employers do not measure how much is being spent on contingent labour and whether it is effectively spent, and 85% of employers do not measure the productivity of contingent workers.[45]

THE CONTINGENT WORKFORCE – BARRIERS TO ENTRY

In a presentation by EY, *The Contingent Workforce – Traps to avoid*, it was claimed that HR professionals may not always be aware of the risks and issues they may expose the organisation to when hiring contingent employees, and that not all organisations have defined policies and procedures to mitigate risks when hiring contingent employees.[46]

Although many organisations are moving to contingent models, a key barrier to entry into the contingent workforce model often expressed by HR professionals in organisations is in-country labour laws that may prevent the procurement of services in this manner. In addition, organisations are concerned about privacy legislation, loss of intellectual property, lack of confidentiality and possible brand damage.

The key to an effective contingent programme is that of planning, defined approaches and projects. It cannot be as ad hoc and unplanned as initially believed or implemented. In order to anticipate risks and barriers and still attain the benefits of cost saving, an elastic workforce and access to a pool of skills, the organisation should:

- define the relationship clearly upfront and have consistent approaches to this definition;
- have standard and mandatary NDAs that contingent workers are required to sign at the start of each engagement;
- know and research their contingent workers as well as they would a core employee through reference checks, online ratings, CVs, qualification confirmations etc.;
- make use of platforms that support these arrangements which provide clear mechanisms for on-boarding a worker onto a project and off-boarding them once the project is complete – the cost is worth it; and
- account for the use of contingent workers as clearly as hiring permanent core employees – the management of costs and an analysis of returns should be a requirement.[47]

THE CONTINGENT CULTURE

A question often considered when hiring a contingent workforce is whether they should be engaged with the company culture or not. Although the contingent worker is likely to work for more than one employer, is it possible for them to still feel engaged with the values and culture of the client they are working for? In the OCG study it was found that 86% of employers feel they are giving contingent workers the same treatment as permanent employees, however 19% of contingent workers feel they are not engaged with the company culture. This shows that employers are perhaps not doing enough to make contingent workers feel at home in the company, and should involve them more in the company culture outside of just working. Despite this, the long-held perception that contractors have a negative impact on company culture seems to be a little wide of the mark. Only 4.6% of employers feel this is the case, and 92.5% of employees believe that contingent work is becoming more accepted. This bodes well for further growth of the contingent workforce in the future, as contingent workers become more accepted by both employees and employers alike.[48]

Incorporating contingent workers into a culture is a mechanism whereby the relationship continues beyond "the gig". By incorporating the contingent worker into their culture, employers build a relationship and retain the benefits of having a pool of talented, competent people who understand the values of the business to hand.

CONCLUSION

There is an abundance of literature relating to the benefits, risks and rise of the contingent workforce. The move to this model is rapid in some regions and minimal in others. As organisations and people employed in those organisations find themselves facing the challenges of an uncertain economy, there is a need not only for organisations to embrace the new way of working, but the employees as well.

CHAPTER 3

THE POWER OF PURPOSE – THE REAL LANGUAGE OF RETENTION

Morag Phillips

INTRODUCTION

Are we brave enough to ask our key employees: "What do you LOVE about working here?" Can we even answer that question for ourselves? I would suggest that if there is anything to LOVE about the work we do, it is likely something we feel passionate about, something we feel committed to, connected with, something we believe in. If we feel that way about work, or even just a part of the work we do, we are probably giving our all to it, unaware of the hours we work or the sweat and tears we give. As business leaders we want employees like this in our teams, where they are loving what they are doing (at least most of the time!) and giving their all. As employers, we would give our all in return, and would be demonstrating commitment back to the employee. What is the magic ingredient that makes this all happen? Let's consider that a sense of purpose in work is the foundation element that can turn an employee from being "employed" to being "engaged".

ENGAGEMENT – MOVING BEYOND ATTRACTION AND RETENTION

Attraction and retention – familiar words that remain important in both our daily employee experiences and our long-term strategy development. It seems that these two small words can have a huge impact on our business, yet there is still no perfect solution to achieve these goals. I am intrigued that these two small words remain part of conversations, conferences, books, webinars, articles and the like, even after many years. It is clear that we are still learning!

As a start, let's consider the components of total reward, i.e. the basics of what we should have in place to attract and retain talent. Typically these include a number of pillars, each with a primary and ideal outcome (e.g. attraction/retention), all of which link to an overall organisational strategy:

Organisation Strategy

HR Strategy

Employee Value Proposition (EVP)

This is what makes our organisation the preferred employer
5 Important Pillars
(adapted specifically for you)

Remuneration	Performance Feedback	Career and Development	Work Environment	Inspirational Leadership
• Tickets to the game • Has to be right • Flexible • Internal and External equity	• How am I do-ing? • Development • Control over performance • Link to pay	• I know where I am going • Growth of portable skills • Vertical and horizontal	• Stimulating • Tele–commuting • Work life balance • HR policies	• Leading and managing • Training • Development • Dual career paths
Attract	Motivate	Retain	Enjoy	Enthuse Inspire

Engagement Index (EI) – how do we compare to the National Index?

Figure 3.1: An illustration of an employee value proposition[49]

These pillars describe the components of what we can offer employees in the hope that they drive the right behaviour and retain the key people. The magic ingredient sits in the last pillar of inspirational leadership, and you'll read in this chapter how we as leaders can maximise the power of purpose to drive true engagement.

Attraction and retention of the best talent remains a key business priority. These are two human resource interventions that have moved to a strategic level, as we have realised the consequences of poor attraction and poor retention of good skills. In the world today, it is probably fair to say that many employees are in a job because of what it can give them, not because of their loyalty to the organisation, which may have been a factor in the 1970s. Employees demonstrate their freedom of movement between companies and opportunities, and continue to move freely even when they have received a lot from an organisation. The shift here is that the organisation cannot assume that excellent employees will choose them – they need to create a reason for the choice. There needs to be a connection between the employer and the prospective employee. In a sense, your ability to attract an employee starts way before your recruitment process and placing of an advert – your brand as an employer is what will produce responses to the advert.

There are a number of definitions of attraction and retention that can frame our thinking. The relationship between employee **attraction** and organisational factors can be perceived through the workplace attraction model developed by Amundson. In developing his model, Amundson reviewed different approaches and identified ten attractors that appear to heavily influence workplace attraction:[50]

- Security
- Location
- Relationships
- Recognition
- Contribution
- Work fit
- Flexibility
- Learning
- Responsibility
- Innovation

It is interesting to note that these attraction factors are still relevant in our current workspace, but have been expanded.

Retention, on the other hand, is "an effort by a business to maintain a working environment which supports current staff in remaining with the company".[51]

Bartlett and Ghoshal[52] described the 'war for talent' as competing for talented and skilled workers by attracting them to work for the organisation as well as retaining their loyalty.

Once we have attracted employees, how do we engage them? It is possible that employees can be satisfied (even happy) with their jobs, but are not **engaged**. In other words, if asked about retention, we would tick the box. However, the major difference is to translate the retention and the employee happiness into performance.

Bridger defined engagement as "the extent to which people are personally involved in the success of a business".[53] This proposes that there is a combination of business success (performance as an output) and personal connection (an emotional response); if we are emotionally connected to something, we are likely to be committed to it.

Here are some more lessons on the concept of engagement:

Engaged employees stay for what they **give** while disengaged employees stay for what they **get** from you.[54] Again, this demonstrates that an employee may stay (they are being retained), but they are not engaged to the point where they focus on the "give".

Gallup described that "engaged workers stand apart from their non-engaged and actively disengaged counterparts because of the discretionary effort they consistently bring to their roles. These employees willingly go the extra mile, work with passion, and feel a profound connection to their company. They are the people who will drive innovation and move your business forward".[55]

As Franz[56] noted, employee engagement is *not*:

- a strategy;
- a mandate;
- employee motivation;
- employee recognition;
- something that is "done";
- an organisational competence;
- a morale booster;
- a performance booster;
- performance goals;
- a reward programme;
- an investment;
- an incentive;
- a survey;
- trainable;
- coached;
- a training programme;
- technology-driven;
- a management style;
- a party every Friday afternoon;
- unlimited free food and similar perks;
- a plaque on the wall;

- a shirt with your logo on it;
- education reimbursement;
- employee satisfaction; and/or
- employee happiness.

Franz shared a simple equation that shows the confluence of passion and purpose:[57]

I believe in
the brand

I believe in
what the
organisation
does

I am
engaged

Figure 3.2: Passion and purpose

Engaged employees believe in their future and invest in it. They become active participants rather than judges on the side-line, and they see that they are valued. By implication, if an engaged employee demonstrates commitment and investment, the result should be positive. A number of studies demonstrate the direct link between employee engagement and bottom line business success.

The key question, then, is how do we engage employees? Does money fill that need? We have empirical evidence that supports the view that money can have an impact as an initial motivator; it can attract an employee and does certainly have an impact on an initial decision to join an organisation. It has been called a "hygiene factor", implying that if the organisation "keeps things clean" and manages a robust remuneration structure, pay dissatisfaction will not creep in. However, once the money part of the transaction is done, it becomes invisible and we need to build a deeper engagement driver. Along comes the exciting concept of a purpose-driven organisation!

Let's refer to an inspirational model introduced by Sinek and examine how he has broadened our perspective to consider WHY. In his view, "If you hire people who believe what you believe, they'll work for you with blood and sweat and tears".[58]

Sinek's concept is that if you start with 'Why?', this creates the foundation which establishes the cause, purpose and belief of the organisation, as well as why it exists.[59] Starting with 'Why?' is the hallmark of a successful individual. The thinking is based in biology. Our neo-cortex is responsible for our rational and analytical thought and language – the how and what. Our limbic brain is responsible for feelings – the why. When we feel a response it has an impact and it can influence what we choose to be committed to. Purpose driven organisations start with why.

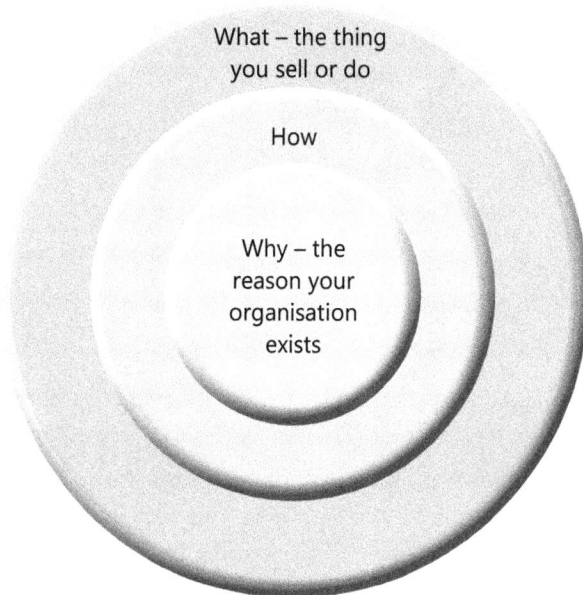

What – the thing you sell or do

How

Why – the reason your organisation exists

Figure 3.3: Golden Circle Model[60]

Think of the famous speech by Martin Luther King Jr... "I have a dream..." not "I have a plan"! He sowed the seed of the purpose, the feeling, the WHY. What is the WHY in your organisation? Once you have a WHY, the 'how' flows out in the form of the culture, the processes and the way things happen in your organisation.

As suggested by Roshan Thiran, "...there is a difference between giving direction and giving directions. Direction is the end destination (your WHY)... whilst directions (your HOW and WHAT) is the plan to get you there".[61]

Pontefract expanded this thinking:[62]

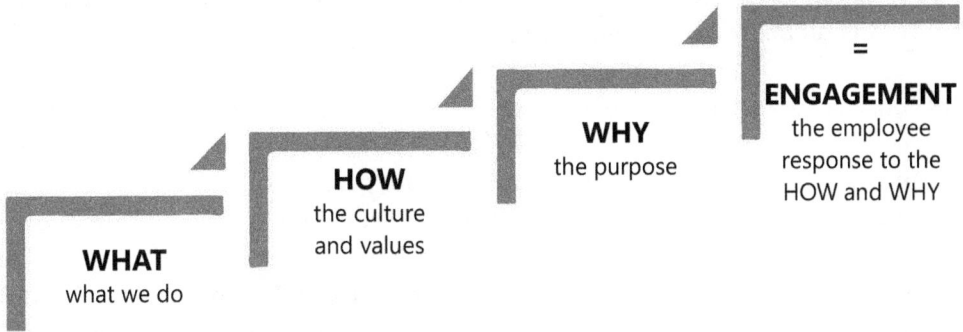

Figure 3.4: The Why-How-What model[63]

Thus engagement is about both the 'how' (the culture) **and** the 'why' (the purpose). With these models in mind, engagement may be impossible without a clear purpose.

WHAT ARE THE BENEFITS OF ENGAGEMENT?

Table 3.1: Benefits of engagement

INCREASED	REDUCED
Productivity	Recruitment costs
Innovation	Employee turnover
Commitment	Negative culture
Customer satisfaction and loyalty	Mismatched employees
Discretionary effort	
Profits	
Retention	

We see people choosing passion and a cause over money. Studies suggest that millennials are a purpose-driven generation and they choose careers that give them meaning and an opportunity to make a difference in the world. On the other end of the age spectrum, older people may have had a career of wealth creation and personal growth, and are at a stage in their lives when they wish to offer a contribution back to society.

33

Having a meaningful job is generally associated with doing something that aligns with one's values. If this alignment is there, it will likely also make us more committed to the cause of the organisation we are working in.

Let's consider some definitions of purpose:

- An aspiration. This implies that it has a long-term outlook.
- Why do you exist? What does the organisation stand for?
- Purpose is about making a difference and having an impact.
- Purposeful work is about doing something meaningful. This implies that if people value different things, then they will be attracted to different purposes and consequently different organisations.
- An organisation without purpose *manages* people and resources, while an organisation with purpose *mobilises* people and resources.[64]
- A purpose driven company is one that "has an important objective that creates meaningful impact for stakeholders".[65]

CONCLUSION

"Purpose is the foundation on which engagement is built." Biro noted that "one is not better than the other. Without a sense of purpose, employee engagement may be short-lived, and without engagement, a company's purpose will not be fully realised".[66]

Does purpose have a purpose? How does it translate into real business results?

- It drives innovation and change.
- It connects people to their work and solidifies their commitment, both to the organisation and to each other.
- There is a link between engaged employees and great customer service.
- Customer retention translates to business success and can influence profits.
- Strong business returns make for satisfied shareholders who keep the brand strong.
- It retains employees who are engaged. By definition, they are the passionate people who influence good outcomes and spread the contagion of engagement.

A group of Harvard researchers in the 1990s established the following links:

High quality leadership
- Leadership
- Internal service quality
- Employee satisfaction

Enhanced people outcomes
- Employee satisfaction
- Discretionary effort
- Intent to stay
- Productivity
- External value

Enhanced business impact
- Customer satisfaction
- Customer loyalty
- Revenue growth
- Profitability

Figure 3.5: The service-profit chain[67]

In simple terms, an engaged employee translates into business success.

If all of these great outcomes are driven by something as simple as a purpose, it seems critical that we learn how to do that! If an organisation lives only for profit or power, employees work because they have a job, their primary motivation being a salary at the end of the month. Pontefract[68] called this a "job mind-set". Is it easier for a service-oriented profession to say what their purpose is? A doctor, nurse, paramedic, environmentalist, a teacher or a pastor... what about a business professional? Stern[69] reminds us that Peter Drucker had a suggestion: "The purpose of business is to create a customer."

Let's consider a few company examples where a clear purpose has been identified, some of which are well known brands:

ImpERATiVE.

An organisation started by Aaron Hurst, Imperative defines purpose in three categories:[70]

- Creating a positive impact.
- Connecting with other people by building meaningful relationships.
- Achieving continued personal growth.

Google

- A clear mission: to organise the world's information, and make it universally accessible and useful.[71]

"People look for meaning in their work. After all, we spend more time working than we do almost any other activity in our lives. People want all that time to mean something."[72]

Apple

- We believe in challenging the status quo. We believe in thinking differently.[73]

"Your work is going to fill a large part of your life, and the only way to be truly satisfied is to do what you believe is great work. And the only way to do great work is to love what you do".[74]

21st Century

sustainable remuneration in a changing world

- Be the recognised global reward company with happy employees.[75]

Unilever

- Making sustainable living commonplace.[76]

WALT DISNEY

- Promote and spread happiness.[77]

NIKE

- Bring inspiration and innovation to every athlete in the world (P.S. if you have a body you are an athlete).[78]

EY

- Build a better working world.[79]

These examples illustrate a broad range of focus areas and many variations on what may appeal to employees. As these examples demonstrate, a stated purpose does not have to be formal or technical, nor does it have to speak business language. All of these statements have an aspirational element. They all speak to an emotional response – if any of these statements appealed to you, it likely created a feeling. This feeling creates a desire to be associated with the brand and to be known for being part of the team that lives this aspirational goal. It drives commitment and a desire to contribute and perform. Not only does the purpose create a statement of intent, but it also translates into the culture, where the purpose can be felt in the lifeblood of the organisation every single day. What better example of true engagement?

How do you create purpose in an organisation?

- Find your WHY. Treat this as an imperative with your leadership team. It is likely that your WHY is already established and being lived in your organisation. Perhaps it just needs to be articulated and simplified.
- Ask employees what their purpose is at work – find their WHY.
  ◦ What would our customers miss if we were not here?
  ◦ I get up in the morning because...
  ◦ I wish I could...
- Link your organisational WHY to the employee WHY.
- Clearly communicate the WHY – your mission, vision, values.
- Create and drive a culture of honest feedback, courageous conversations, trust and vulnerability.
- Remind yourselves of your organisation's values.
- Communicate your purpose at the recruitment stage – employees are signing up to a purpose, which means you bring in the right people.
- Take care of employees.
- Celebrate success and recognise achievements by employees and teams. Recognition can be a way of giving employees a sense of purpose. Feedback enhances development and is forward-looking – it immediately demonstrates a commitment to an individual's journey.
- Actively move from managing talent to creating an optimal culture where talent can thrive.[80]

As you drive the purpose of the organisation, you will see passion bubbling up in your employees; you will see them starting to behave like owners and participants of a broader mission. This shared purpose can be contagious – a good thing to share! Consider the example from Antoine de Saint-Exupery: "If you want to build a ship, don't drum up the men to gather up the wood, divide the work and give orders. Instead, teach them to yearn for the vast and endless sea."[81]

What does purpose do?

- It unites.
- It attracts the right people who buy in to the same purpose, thereby building a force for good.
- According to Pontefract, purpose does not only serve to engage employees, it also ultimately results in great customer experiences:[82]

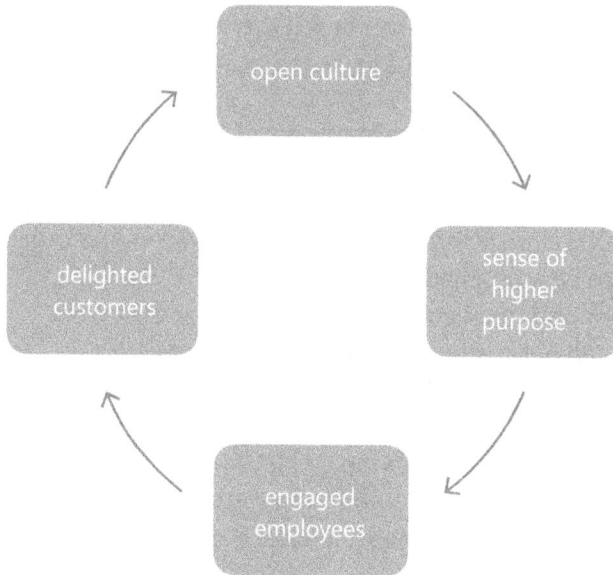

Figure 3.6: Creating purpose in an organisation

This chapter has built a case for building a purpose-driven organisation, so that you are doing more than attracting and retaining talent. Moving beyond attraction, **engagement** makes the most of the employees you are retaining. It is the perfect example of win-win... your employees are happy, they are committed, they perform, they keep your customers happy and your organisation prospers as a result.

Go ahead and take the challenge. Ask your key employees and yourself: "What do you LOVE about the work you do?"'

Enjoy the journey!

CHAPTER 4

THE USE OF AGILE ORGANISATION DESIGN TO DRIVE ALIGNMENT WITH AUTONOMY

Dave van Eeden

INTRODUCTION

According to Veldsman[83], the probable future world of human work will be characterised by a context which he described using the acronym VICCAS: Variety, Interdependency, Complexity, Change, Ambiguity and Seamlessness.

The implications of this emerging world of work on organisation design are profound. This chapter will attempt to address how organisation design is changing in order to address these new environmental and organisation imperatives, so that organisations are able to continuously retain competitive advantage through rapid, flexible adaptiveness, without a loss of rigour and quality performance. There are also significant implications for many other aspects of organisational life, not least that of leadership, which will be mentioned as they arise but are not central to the chapter and are addressed elsewhere in the book.

ORGANISATIONS AS OPEN SYSTEMS

The history of thought about organisations describes, as far back as the 1950s and 1960s, the view of organisations as open systems that have to cope with the influence of the environment. Morgan[84] described the open systems approach to organisations and the impact of several key issues on the organisation as such:

- The role of the environment and its impact on the organisation because the organisation is not a closed system.
- The organisation is a set of interrelated sub systems.
- There needs to be alignment between different systems in the organisation to minimise dysfunctionality.

This approach became known as the contingency approach to organisation, which is underpinned by the following biological system-based concepts:

- An open system emphasising the interaction between the system and its environment.
- Homeostasis, referring to self-regulation by the system and the maintenance of a steady state.
- Entropy and negative entropy, which describe how closed systems deteriorate and run down while open systems attempt to sustain themselves by importing energy, thus displacing negative entropy.
- Structure, function, differentiation and integration, emphasising the non-linear interdependence and interconnectedness of all parts of the system and cautioning against reductionist, linear, and cause and effect organisation analyses.
- Requisite variety, stating that the internal regulation of the organisational system must be at least as diverse as the environment it is attempting to cope with, in order to adequately respond to that environment.
- Equifinality, which emphasises the flexible patterns of organisations inherent in a living system, as opposed to the fixed cause and effect system relations in a closed machine system.
- System evolution, which describes the capacity of an open system to evolve to more complex forms of differentiation and integration, thereby enhancing the ability of the organisation to cope.[85]

Based on the work of Burns and Stalker[86], Morgan contrasted, for the sake of illustration, how different organisations respond to various dimensions relative to the system view of the organisation.

This is shown in summarised form in Table 4.1.[87]

Table 4.1: Organisation responses to dimensions

Dimension	A Rayon Textile Mill	An Electronics Firm
Nature of environment	Relatively stable, technological and market conditions relatively well understood	Highly unpredictable, rapid technological advances and boundless market opportunities
Nature of task facing the firm	Efficient production of standard product	Exploitation of rapid technical change through innovation and exploration of new market situations
Organisation of work	Clearly defined jobs arranged in hierarchical pattern	Deliberate attempt to avoid specifying individual tasks; jobs defined by the individuals concerned through interactions with others
Nature of authority	Clearly defined and vested in formal positions in hierarchy; seniority important	Pattern of authority informal and constantly changing as roles become redefined with changing circumstances; vested in individuals with appropriate skills and abilities
Communications system	According to a pattern specified in various rules and regulations; mainly vertical	Completely free and informal; the process of communication is unending and central to the concept of organisation
Nature of employee commitment	Commitment to responsibilities associated with their own particular jobs; loyalty and obedience important	Full commitment to the central tasks facing the concern as a whole and an ability to deal with considerable stress and uncertainty
Systemic tendency	Mechanistic	Organic

THE TRADITIONAL VIEW OF ORGANISATION DESIGN

In spite of the above history of thought, the fundamental paradigms dominating traditional, and to a fairly large degree current, practice and thinking on organisation design can be described in three dimensions:

First, the environmental view assumes a relatively orderly, predictable environment that allows adequate response time and competitive advantage based on a relatively stable set of parameters.

The second dimension is a view of the human worker that is based on Douglas McGregor's[88] Theory X and Theory Y. The prevailing view was that of Theory X; that a worker is not really concerned about quality, excellent performance and a will to produce. The implication for organisation design in this dimension is that control and structure are needed to ensure efficiency and effectiveness. This is best described by Thoren[89], where Human View X is characterised by the view that:

*   people dislike work, find it boring, and will avoid it if they can;
*   people must be forced or bribed to make the right effort;
*   people would rather be directed than accept responsibility (which they avoid);
*   people are motivated mainly by money and fear about job security; and
*   most people have very little creativity – except when it comes to getting around the rules.

Human View Y is characterised by the view that:

*   people need to work and want to take an interest in it. Under the right conditions, they enjoy it;
*   people will direct themselves toward a target that they accept;
*   people will seek and accept responsibility under the right conditions;
*   under the right conditions, people are motivated by the desire to realise their own potential; and
*   creativity and ingenuity are widely distributed and grossly underused.

The third dimension is a paradigm of viewing the organisation as a 'machine', with what Morgan[90] described as a set of mechanical relations – a state of orderly relations between clearly defined parts that have some determinate order. This machine model is perhaps best known by the work of Frederick Taylor[91] and the concept of scientific management, such that the period 1911-2011 was described as the "management century".[92]

These three paradigms have had a pervasive impact on organisation structure and design, such that organisations have failed either because of a lack of response to a changing environment or because of a Theory X view of people which has within it its own self-fulfilling prophecy (if I have a Theory X view of people they will behave that way); an inflexible, rigid cause and effect view; or a combination of the two.

This, it is contended, has largely taken place in spite of the contingency view of organisation structure described above. Contingency approaches to organisation design, while tacitly acknowledging the need to respond to the environment organically, have largely failed to fundamentally make this transition.

AN EMERGING MODEL OF ORGANISATION DESIGN

In recent years there has been a shift from the rigidities of the above three traditional paradigms. More enlightened and forward thinking organisations and leaders have started considering the value of people as the only source of competitive advantage, in part because the importance of collaboration in teams began to be emphasised and because large, often multi-national, organisations have tried to cope with an increasingly VICCAS world through more appropriate structures and systems. Yet it is still contended, perhaps cynically, that these approaches, although well intentioned, are still largely based on a Theory X view of people, a machine view of organisations, and a largely command and control, power-based view of management and leadership, albeit couched in a participative, authentic leadership philosophy.

In addition to these shifts, McKinsey identified four disruptive trends challenging the traditional organisation structure paradigm in business and industry.[93] These are:

- a quickly evolving environment with rapidly evolving stakeholder demands;
- constant introduction of disruptive technology, resulting in the commoditisation of previously established businesses and industries and the digitisation and application of technology to these businesses and industries;
- accelerating the digitisation and democratisation of information. This increased volume, transparency and distribution of information results in rapid engagement in multi-directional collaboration and communication; and
- the new war for talent requiring 'learning' workers, with the resultant emphasis on a different employee value proposition to attract and retain diverse and different workplace demands.

In essence, McKinsey is predicting an organisation paradigm that will need to meet contradicting demands of stability and dynamism, posited on a move from the traditional hierarchical, bureaucratic, silo-based structure (which provided stability but is not coping with the emerging disruptive environment) toward an organisation design based on quick responses, flexible resource allocation, a focus on action (not structure and reporting lines) via teams with end to end accountability, and most importantly, leadership providing direction and an enabling culture. This new organisation has been termed the 'Agile organisation', characterised by a move from 'organisation as machine' to 'organisation as organism'. The McKinsey definition is worth repeating:

"An Agile organisation is a network of teams within a people-centered culture, that operates in rapid learning and fast decision cycles, which are enabled by technology, and that is guided by a powerful common purpose to co-create value for all stakeholders."[94]

Pia-Maria Thoren, in her book *Agile People*[95], contrasts the traditional and Agile organisations as follows:

Table 4.2: Traditional to Agile

From – To	Traditional	Agile
Processes	Episodic	Ongoing
	One size fits all	No size fits all
	Standardised	On a needs basis
	Reactive	Proactive
	Push	Pull
Organisation	Machine	Network
	Individual	Team
Leadership	Management	Employeeship
Human view	Negative (Theory X)	Positive (Theory Y)
Motivation	Extrinsic	Intrinsic
Feedback	Seldom	Often
Human resource role	Controls, implement standards	Support and coach organisation agility

AGILE ORGANISATION DESIGN – A PANACEA?

The field of practice in management and especially Human Resource management (inclusive of organisation development and organisation design) is replete with fads, quick fix solutions and solutions to problems and challenges, which have not necessarily been clearly thought through or executed fully.

In order to avoid Agile organisation design and structuring falling into this trap, it is necessary to interrogate under what circumstances Agile Organisation Design would be the most suitable approach, as opposed to engaging in an expensive and complex change process because this is the supposed new way of structuring a business. The table below attempts to provide a check list for consideration before implementing an Agile structure.

Table 4.3: Agile organisation design considerations

Dimension	Consideration	Agile Design is suitable in conditions:
Environment	Volatility	of high and often unpredictable environmental validity;
	Interdependence	where interdependence of task and role input and output are reciprocal through the value chain;
	Complexity	where problems and tasks are complex with multiple inputs and no clear solutions;
	Amount and speed of change	where change is rapid and endemic and where speed of response is a source of competitive advantage;
	Ambiguity	where tasks contain inherent ambiguity and an undefined scope;
Organisation Vision	Does the organisation vision embrace audacious future aspirations and dreams?	where the vision is audacious, complex, never been realised before and requires unique beliefs and skills;
Organisation Mission	Does the mission of the organisation require management of a complex, changing value chain?	where constant supplier collaboration; ongoing, changing customer demands; and iterative internal conversion processes exist;
Organisation culture and values	Is achievement of organisation vision and mission dependent on cross functional collaboration and employee engagement throughout the value chain, including customers and suppliers?	where trust in people, collaborative working relationships and highly engaged and committed employees are needed;
	Is cutting edge quick response innovation required?	where innovation is a pre-condition for success and where ongoing development is essential;
	Is a learning culture required to maintain competitive advantage?	where a requirement for success is the development and maintenance of a learning culture characterised by a premium being placed on continuous learning and coaching;

Dimension	Consideration	Agile Design is suitable in conditions:
Leadership	What is the view of people held by leadership and what does leadership view their role to be?	where a holistic view of people and their contribution to the enterprise is critical;
Architecture and resources	What processes and systems are required to ensure competitive advantage?	where quick response and self-adapting processes and systems are appropriate for the nature of the business;
	What are the work demands of the talent that needs to be attracted and retained to achieve competitive advantage?	where the talent required to achieve the vision and mission of the business is diverse and demands flexibility, the use of technology and collaborative work practices; and
	Is the organisation mission dependent on staying abreast of technology?	where cutting edge technology is a pre-requisite for business success.

Bain and company[96] have also suggested guidelines for adopting an Agile culture with the following areas of consideration:

- Where the market environment is such that customer preferences and solutions change quickly.
- Where collaboration and rapid feedback are feasible and where customers become clearer on their needs as the process progresses.
- Where innovation is required due to complex problems, unknown solutions, undefined scope and where product specifications may change and time to market is important.
- Where the modularity of work is high; where incremental development is high, value adding and useable; and where work is iterative, cyclical and can be broken into parts.
- Where the impact of mistakes is not costly but rather a learning opportunity.

Given these criteria and conditions, and being mindful of the need to implement organisation design to address real organisational design needs, what is emerging is what Ulrich has termed a new organisation capability – what an organisation is known for and good at doing to deliver value to its key stakeholders. He named this capability "agility", i.e. the ability to anticipate and/or quickly respond to emerging market opportunities. Ulrich posited that agility matters at four levels:[97]

- Strategic agility – an ability to build capacity for continual strategic change, moving from:
  - industry expert to industry leader;
  - market share to market opportunity;
  - who we are to how customers respond to us;
  - existing markets to new and uncontested markets;
  - beating competition to re-defining competition; and
  - blueprints for action to dynamic processes for Agile choices.
- Organisation agility – enabling an organisation to anticipate and rapidly respond to changing market conditions. This includes creating autonomous market-focused teams that can move rapidly to create and define new opportunities.
- Individual agility – the ability of people and leaders to learn and grow, consisting of a mind-set and a set of skills that enable individuals to change as fast as their work demands.
- Human Resource agility – people systems and practices which foster agility but, even more importantly, create an Agile organisation where autonomous teams are connected to other teams in an independent organisational eco system.

A FUTURE POSSIBLE ORGANISATION MODEL

McKinsey provide some thoughts on the possible organisation of the future.[98] What they envisage is what they term "Organizing for Urgency", which contains the dimensions listed below in Table 4.4. The implications for organisation design in these dimensions have been added in the third column:

Table 4.4: The possible organisation of the future

Focus	Action	Key success factors
Urgency	An absolute focus on speed – making decisions with 70% of the information	A culture which permits mistakes
Emergent strategy	A shift to emergent strategy – the relentless pursuit of value creation is the strategy	Absolute clarity of vision and mission
Agility	Encouraging real time decision making throughout the organisation	Competence to make decisions and an encouraging culture
	A flat organisation structure decoupling rank and title from control	Leadership which understands and lives this approach

Focus	Action	Key success factors
Capability	Personalised talent programmes using advanced analytics	A sophisticated talent strategy
	A leadership model which acknowledges that leadership can stem from anyone and is earned not appointed	Leadership which understands and lives this approach and a culture which reinforces it
Identity	The adoption of a recipe that permits the selection of the best approach to create value	An understanding and formulation of an approach to strategy which encourages order without chaos and creates value
	The cultivation of purpose, values and social connection so that individuals in the organisation are aligned around principles	Absolute strategic clarity of vision and mission

THE CURRENT SITUATION

The status quo seems to be as follows:

- The future world of work will be characterised by a VICCAS world.[99]
- The history of management thinking has posited the open-systems approach to organisation design since the mid-twentieth century based on the principles of the organisation as an organism not a machine.
- Three paradigms still tend to underpin the traditional view of organisation design – a predictable environment, a Theory X view of people, and a mechanistic view of organisations.
- Organisation design theory, while conceptually recognising the need for an organic approach, is still largely based on the command and control model.
- A new model of organisation design is emerging triggered by four severely disruptive trends, necessitating a shift to 'Agile organisations'.
- Agile organisation design is not a panacea and should be considered to address several organisation design imperatives to support strategy execution.
- Agility is emerging as a new organisation capability.
- New organisation designs, based on an absolute imperative for urgency, are emerging.

KEY CONSIDERATIONS IN THE DESIGN OF AGILE ORGANISATIONS

The next question that must be addressed is what the key considerations are in the design of effective and sustainable Agile organisations, so that stability and agility are balanced.

In considering this question, guidelines were drawn from the literature on this subject using a generic model of organisation effectiveness, encompassing the:

- purpose, vision and clarity of direction of the organisation;
- mission of the organisation – what business is it in and who are its customers?;
- leadership standards and behaviours;
- work teams; and
- measurement and review.

These five dimensions of effectiveness are enabled and supported by:

- culture;
- technology;
- processes and systems;
- talent; and
- learning and people development.

Key considerations and factors are noted and described under each of the above ten dimensions – these can be regarded as key success factors or as guidelines to consider in Agile Organisation design. Reference is made to the work of McKinsey in this regard.[100]

- Purpose, vision, clarity of direction:
  - A clear, shared purpose and vision throughout the organisation enables stability, passion for stakeholders, and the facilitation of proactive anticipation of stakeholder needs.
- Mission of the organisation:
  - Absolute clarity on the business of the organisation and its stakeholders.
  - Fostering an emergent mix of multiple strategies in support of the vision.
  - Flexible resource allocation in support of the mission and the ability to quickly change resource allocation between initiatives, based on regular review, in support of the mission.

- Leadership standards and behaviours:
  - Immediate actionable strategic guidance to work teams by leaders, who play an integrative role in ensuring tangible value delivery.
  - Provision of ongoing feedback and coaching.
  - Facilitation and support of action-oriented decision making.
  - Creation of a cohesive community with cultural norms reinforced through positive peer behaviour, rather than rules or hierarchical power.
  - Shared and servant leadership committed to and competent at being visionaries, architects and coaches.
  - Fostering of an entrepreneurial spirit.
  - Leading by influence.
- Work teams:
  - A stable top-level structure.
  - A clear flat structure.
  - Clear accountable roles.
  - Role mobility which allows for shared and multiple roles.
  - Facilitation of active partnerships.
  - A flexible, scalable, networked organisational eco-system both externally and internally.
  - Fit-for-purpose work cells designed to deliver clear value-adding outcomes.
  - Removal of unnecessary hierarchy.
  - An open physical and virtual environment which encourages transparency, communication and collaboration.
- Measurement and review:
  - Hands on, transparent governance, decision-making and review.
  - Continuous feedback, review and open discussions of performance against targets.
  - Complete transparency and ease of access to information.
- Culture:
  - A culture where people are authentically placed at the centre, engaging and empowering everyone in the organisation.
  - A culture of sensing and seizing opportunities in support of the purpose and mission.
  - A performance-orientated culture.
  - A culture of transparency.
  - High standards of alignment, accountability, expertise, transparency and collaboration.
  - A culture of quick, efficient and continuous decision-making underpinned by a principle of 'disagree and commit'.

- Technology:
  - Continuously evolving architecture, systems and tools.
  - Development of next generation technology and tools.
- Processes and systems:
  - Rapid iteration and experimentation.
  - Standardised, stable ways of working to encourage and facilitate interaction and communication.
  - Disciplined application of processes and systems.
  - Ruthless removal of unnecessary bureaucracy.
- Learning and talent:
  - Encouraging the nurturing and development of inspiring individuals.
  - Continuous learning and a focus on continuous improvement.
  - Encouragement of robust functional communities of practice which serve as professional homes for people as they fulfil different roles.
  - An open talent marketplace which supports people development.

CRITICAL SUCCESS FACTORS

From the above review of the literature and experience on the emerging Agile organisation design model, the following are postulated as being critical success factors in successfully designing Agile organisations while also retaining necessary stability: agility is critical for dealing with the turbulent environment and market place, and requires speed, responsiveness and adaptiveness; stability is needed to ensure reliability, efficiency, quality and governance.

These two contradictory imperatives can be explained through a music metaphor – what is optimal and desirable is 16 bar jazz with the optimal mix of order and creativity, as opposed to the extremes of disorderly 32 bar jazz or the conservatism of 8 bar jazz.
To achieve this, and before embarking on an Agile organisation design of this magnitude, the following eight factors are deemed to be critical:

- Authentic leadership commitment to people and the concept of leadership through coaching, influence and integration.
- Absolute clarity of purpose and vision.
- Visible and transparent commitment to defined cultural norms and practices.
- Ruthless pursuit of the disciplined execution of stable ways of work.
- Rapid proactive response to market and customer needs with flexible resource allocation.

- Detailed attention to the many technical aspects of organisation design: decision-making, team structures, team behaviour and focused implementation of the eco-system.
- Enabling the use of technology.
- Visible people and talent development.

CONCLUSION

In conclusion: "New science requires us to question many of our most deeply held assumptions about how things work in life and in our organizations. None of these shifts is insignificant. All of them are worthy of further thought and conversation, as we try to invent and discover the organizations of the next century. Hopefully these newer sciences point the way to a simpler way to lead organizations. But to arrive at that simplicity, we will have to change our behaviours and beliefs about information, relationships, control and chaos. We will need to recognize that we live in a universe that is ordered in ways we never suspected, and by processes that are invisible except for their effects".[101]

⌕ CASE STUDY   **The SEMCO Example[102]**

Introduction

The SEMCO organisation is worth describing as what could be termed an 'extreme' example of an Agile organisation design.

Described in the books *Maverick*[103] and *Seven Day Weekend*[104] by Ricardo Semler, the SEMCO story provides an excellent illustration of what would be required to make a successful transition to an Agile organisation design.

The SEMCO Style Institute describes the history of SEMCO Style as follows:

In brief, until 1980, SEMCO was run by its founder, Antonio Curt Semler, in Brazil. SEMCO Style came into being when 21-year old Ricardo Semler, a law student, took over the SEMCO Group, a centrifuge manufacturer, from his father. On his first day, the young Semler fired 60% of all top managers. He continued to democratise the company, turning the old corporate hierarchy on its head by delegating as much decision power to the workforce as possible. He also expanded the Group's offering by moving heavily into the service sector, including environmental consultancy, facilities management, real estate brokerage, and inventory support.[105]

In 20 years he grew the company from 90 to 3,000 workers, and from $4 million dollars to $212 million dollars in annual revenue – with an employee turnover of only 2%.

SEMCO Group has since evolved into SEMCO Partners, which works with a variety of technology partners through a sophisticated joint venture model that combines SEMCO management practices with the partners' expertise and product lines. This fusion of skill sets has created a business model that has been extremely successful in the Brazilian market.

From 1953-1979, the organisation was a mechanistic organisation characterised by hierarchy, command and control structures, power games, little creativity, and minimal access to confidential company data. "It was a familiar company, autocratically styled – a place where the powerful ruled and the wise obeyed." After 1980 when Ricardo Semler began running the company, and as a result of a range of daring organisation interventions on his part, the business can now be described as a truly organic organisation.

The process

In hindsight, the key practices that were responsible for transforming SEMCO can be described as democratic, participative and sensible.

These practices evolved through a process of reflection among the Board of Directors and company management, resulting in the establishment of a set of long-term goals as well as the values, principles and characteristics that they wished SEMCO to embody. They also realised at the time that they needed to involve all people in the organisation in the crafting of the future SEMCO, as not to do so would nullify the principles they were trying to inculcate, in that they would be continuing to treat people as immature, irresponsible and continually needing to be monitored, controlled and given instructions.

Consequently, a series of actions took place which resulted in what could be described as a revolutionary organisation transformation over a period of several years.

These initiatives included the following:

- Change first; discuss later – they created of a set of conditions which made it inevitable for everyone to be involved immediately, thus avoiding incrementalism, trial and error and top-down change processes. These changes included the distribution of decision-making power, access to all information at all levels, and profit sharing by all employees.
- Vest people with decision-making power – they encouraged staff to make decisions on work and related matters that affected them, rather than referring matters to a supervisor or manager.

- Do away with information secrecy and be open – they created multiple channels of communication and there was no confidential or reserved information.
- Educate employees to interpret all types of information – information transparency was accompanied by a deeper knowledge of the meaning and import of the information.
- Decide profit-sharing – a participative process was created so that through wide discussion and debate, a set of profit-sharing principles was established that organically facilitated and encouraged active participation in the organisation's destiny.
- Expect resistance from management and acknowledge that, for a variety of reasons, management will resist these changes – this required empathy and discussion, as it necessitated stepping out of strong comfort zones created by years of past practice.

A set of 'do's and 'don'ts' were developed after the execution of this process, which are summarised below:

DO

- set goals for the company that are complementary to the goals people set for themselves;
- show people how to think of work as a source of accomplishment rather than an obligation;
- strive to preserve people's dignity and resolve issues in a way that promotes growth on both sides, even when there's a conflict;
- make it safe for people to experiment and fail and to convert failures into teaching moments;
- judge people by their personal qualifications and not by their place on the company hierarchy;
- encourage people's quest for knowledge and allow them to seek answers from anyone with the right knowledge or experience;
- hold everyone accountable for results when everybody participates; and
- approach management in a people-centric way that motivates personal development if you want your employees to discover the best in each other and solve problems creatively.

DON'T

- constantly check-up on people's progress if you want them to be their most productive;
- think of it as a slight if a team member seeks the help of someone else to solve a problem;
- worry that you will lose face if you do not know the answer to a problem – rather convert it into an opportunity to learn and grow;

- punish everybody because a few exceptions abused their freedom or power;
- expect people to trust you if you cannot trust them with absolute transparency; and
- quit trying to transform yourself from an authoritarian and highly hierarchical organisation, even though it is bound to be tough for your company to change.

Primary principles of the SEMCO Style

- Participation – everybody has the right to:
  - get and supply information regarding everything that's going on, from parking-related issues to the company's results;
  - take part in small and big decisions, from the choice of uniforms to the choice of future leaders; and
  - take part in the company's profits, with full understanding of the numbers and allocation of shares.
- Seriousness and reliability.
- Honesty and transparency.
- Equal opportunity.
- Freedom of expression/opinion.
- More autonomy to people, less control and more responsibility; control should be over processes and not over people.
- Search for sustainable results.
- Responsibility to the customer comes before profit.
- A dignified and respectful treatment of people irrespective of their level (hierarchical, social, economic, schooling).
- Work is not just an obligation. It can also be a source of self-realisation, informality, relaxation, naturalness, joy.
- Absence of prejudice; there is no discrimination based on sex, age, race, origin or appearance.
- Fewer hierarchical levels.
- Freedom to make direct contact with any person in the organisation (of any level and/or any company of the group).
- Bold people are motivated to dare and even run the risk of making mistakes; mistakes are regarded as opportunities for learning and growth.

CASE STUDY — Semenco Journey

This case study serves to provide a simple example of this journey:

Clovis Bojikian from the Human Resources department at SEMCO was present when the company began taking steps to adopt participative leadership practices. Recalling the time, he says there was a genuine interest in transforming the workplace from being a place of suffering and obligation into one of satisfaction, joy and accomplishment. In other words, into a place where people could be happy working. "The workplace didn't have to be a place where people suffered during the hours they spent there", he said.[106]

Bojikian remembers Ricardo Semler pointing out something that he'd noticed as well: "People come here sort of crushed – as though they are carrying some weight. They sit at a table and keep performing tasks, counting the minutes until lunch time; and then counting the minutes until it's time for them to leave. And when they leave, it's over. I don't feel any vibe. I suppose these people have much more to give than what they are giving. So, we need to discover ways to make this situation change and to reverse it", Ricardo observed.

According to Bojikian, it is not possible for someone to motivate someone else. Instead, each person needs to be motivated on their own. "It has to come from within if it is to last", he said. Although it is not possible to motivate others, SEMCO realised that it was possible to create the conditions for people to feel motivated. "So, we decided to establish the following: let's enjoy everyday situations and try to exercise participation. Let's start with the simple things", said Bojikian.

It was important to keep things simple because the company comprised mainly workers who operated its five factories. "So, everything had to be viable, in terms of workers at the factory floor level," he said, adding that they decided to take advantage of simple problems first.

One such "simple problem" was the endless flow of complaints to the HR department about the company cafeteria. Although the HR department promised to resolve the issues and did what it could, the list of complaints seemed never ending. "There were surveys that showed that the most popular dish, Feijoada, was also the least popular dish. Or people complained, on the same day, that the beans were too hard and too soft. What that means is that it is very difficult to please everybody in an eatery," he explained.

The HR department was tired of handling complaints about the cafeteria and it was decided that the next time someone complained about something, the department would no longer promise to fix the issue.

Instead, Bojikian asked the person who came to complain what their suggestion to improve things was. Completely taken aback, the complainant was quite at a loss for words, so Bojikian suggested that they gather a group of colleagues and come up with a plan that they'd implement, were they in charge of running the cafeteria.

After a few days and some discussion, the group came back to Bojikian with a plan, however he asked them to revise their suggestion because it did not include any inputs from the kitchen staff at the cafeteria and could therefore not be implemented.

The group roped in a few kitchen staff members and they came back to Bojikian with a plan that could very well be implemented. When he told them to go ahead, they were once again taken aback. "So, I told them they'd be part of the first cafeteria committee of SEMCO, which will be in operation for a year. And, within a year, we planned to choose another committee," he said.

From then onwards, SEMCO never ran the company cafeteria because every year a new committee was elected.

Other SEMCO examples then emerged:

- Solve cleaning problems
- Choose uniform color
- Choose the type of New Year's Eve party
- Choose how to compensate business days interspersed between a holiday and a weekend
- Suggestions to improve quality
- Suggestions to improve the manufacturing process
- Tips to improve productivity
- Setting own production targets
- Participation in the choice of place of work
- Participation in the elaboration of positions and salary structures
- Participation in the selection process of the future boss
- Participation in the selection process of future peers
- Participation in the evaluation process of the manager
- Participation in small investments prioritisation
- Profit sharing of the company
- Self-control
- Self-determination of wages

SEMCO Style

From these organic developments at SEMCO, a global SEMCO Style Institute has emerged with a logical implementation framework that consists of five pillars, each of which is underpinned by operating principles and practices.

This framework is as follows:

Trust
Treat adults as adults
Practice unfiltered transparency of information
Decrease the power gap dramatically between the board and the shop floor

Alternative controls
Trust is a foundation
Autonomy on when, where and how to work without abdicating compliance and quality
Common sense as a means to simplify and streamline procedures
Joint decision-making as a way to determine the right choices and considerations

Self-management
Enable people to achieve their personal and work goals
Treat self-management as a healthy eco-system
Think small teams of ten people and operational units of around 150 people
Organise self-management around circles, not pyramids

Extreme stakeholder alignment
Alignment between process/project/customer teams and internal and external stakeholders
Consistency of direction and focus – do what you say you will do
High impact perspective from the outside in to the organisation
Finding common interests between stakeholders through engagement

Creative innovation
Provide creative space for healthy discussion and innovation
Think in the box – start small and amplify
Encourage entrepreneurship throughout the organisation

Underpinning the above framework is a number of practices and techniques, effectively a toolkit, which is available for organic implementation in an organisation that has embraced the principles and wishes to embark on the journey.

CHAPTER 5

THE ORGANISATIONAL DEVELOPMENT CHALLENGE OF GOING AGILE

INTRODUCTION

As Agile methodologies and practices become more mainstream, organisational development (ODev) challenges are emerging.[107] ODev reflects the ethos and culture of an organisation and is at the heart of ensuring the organisation is ready and sufficiently resilient for the changes that the new way of work demands. There are similarities between the principle assumptions of Agile and ODev – both see a learning culture as critical to business success, and both emphasise autonomy and self-direction. The challenge which will be discussed in this chapter is how ODev can be used to support Agile methodologies in an organisation.

WHAT IS ORGANISATION DEVELOPMENT AND CAN IT BE AGILE?

The discipline of ODev is defined as a "planned and systematic approach to enabling **sustained** organisation performance through the **involvement of its people**".[108] This includes managing change[109], underpinned by humanistic and democratic principles.[110] French and Bell[111] added to the above, stating that ODev is also centred on looking at the sustainability of the organisation, thinking long term of its needs and how these can be implemented in the most effective way. Importantly, for these authors, this means a focus on the organisation's culture.

At a principle level, the underlying philosophies of ODev and Agile are similar; there is a focus on learning and development, managing change, and valuing team players as autonomous self-directed people. ODev and Agile practices, however, may differ, and the discipline of ODev needs to adapt in order to support an organisation's Agile transformation journey.

Adopting Agile methodologies provides a challenge to ODev. Traditional ODev practices need to be assessed for relevance and utility in an Agile organisation. ODev has been criticised for being outdated, too simplistic, and failing to acknowledge that organisations operate within broader systems and have complicated internal matrices that need to be considered. ODev must consider elements such as existing hierarchy and power structures, which have often existed for a very long time and which were probably successful in their time. Changing the status quo in these instances may be more difficult than starting from scratch and creating a culture that is suitable to an Agile way of working.

Successful organisations are adaptable, anticipate change and align to what the market requires. Organisations are now acknowledging that to remain relevant and competitive they need to pick up the pace and not only adapt to market requirements and changing customer needs, but proactively anticipate these changes and even disrupt markets by creating new customer needs. Ensuring that an organisation's people policies, processes and practices can enable this new way of working is essential.

(CASE STUDY) **Roche: structure and culture**

The global biotech company Roche considers agility to be both a structural and cultural characteristic. In a 2018 interview, the Chief HR officer noted the following manner in which Roche approaches agility. This resonates with how culture is important to Agile methodologies and ODev.

Agile as a mind-set: leadership buy-in and development

- The organisation is 122 years old but the long institutional history did not prove to be a barrier to adopting Agile practices. Roche has a clear purpose, which importantly has an inherent Agile objective in it – "Doing what patients need next". This requires speed and flexibility, but also stability. Everyone is able to move with greater speed and flexibility when they have solid ground to manoeuvre from.

- The Roche values (integrity, courage and passion) and their focus on patients provides them with base stability, upon which flexible and Agile practices can be built.
- Roche have introduced a leadership programme called Kinesis to ensure leaders are on-board and have understood the new ways of thinking and working required.
- The initial target population was approximately 400 leaders and the next layer is approximately 2,500 people.
- Kinesis sessions have a four day duration including theory, sharing, and a very experiential, reflective component, so people can really understand how this thinking, and ultimately Agile behaviours, manifest themselves in the way that leaders solve problems or make decisions.
- The sessions focus on what gets in people's way or consumes more energy than it should; the extent to which an Agile mind-set enables a team to perform at the highest level; and the role the leader plays.

Agile as a rigorous method and commitment to process

- In addition to the adoption of an Agile mind-set, Roche has also rigorously implemented Agile practices. The rigour is seen as an important driver of success – the Agile practices must be properly and consistently implemented.
- Results include much greater speed – activities that usually took 8-12 months are now taking 2-4 weeks.[112]

Agile Methodologies and Organisation Development are Inherently Complementary

The approach and importance given to ODev in an organisation can arguably be seen as a reflection on how that organisation is able to adopt and use an Agile way of working. At its heart, OD is about learning and adapting; it should be reflective and reflexive, and above all it must appreciate the values of the organisation and human development.[113], [114, 115]

Grieves and Redman[116] argued that ODev has six characteristics:

- A methodology based on action research, that is, research which does not wait until the end of the event to **make changes in process and interventions**. It is a constant reflection on changes as they occur and adaptations to incorporate those findings. In essence, the focus is on the activity taking place, and reflecting and evaluating as it happens in order to make incremental improvements throughout.

- Stakeholder engagement in the process is essential and **stakeholder collaboration** more so. This enables buy-in, as well as consideration of any cultural barriers that may exist.

> **Action research can be agile**
>
> McLean described the action research method of ODev as: plan, do, check, act. At the 'plan' stage, ODev looks to see how the organisation can be improved. 'Doing' involves running a trial of the intervention. This is then 'checked' and finally 'action' is taken. Essentially, this is a trial and error approach, with the flexibility to make changes as needed, and being able to respond to the needs of the organisation quickly. The method itself is inherently agile, but needs to be customer-centric and include constant customer feedback loops with a fast turnaround time to be truly agile.[117]

- An awareness and **appreciation of political processes**. Traditional ODev often neglected this aspect of an organisation, but it is essential to be cognisant of what political power exists in an organisation. You may have executive level approval, but without an influential 'champion' who is able to understand and defend organisation changes and learning, little traction will be gained.

- Emphasis on both organisation and individual learning. **Learning is at the heart of ODev**, and one needs to consider that this learning is not just for the individual advancement of knowledge, it is the learning of skills, interactions, collaboration, support of collaborators organisational processes and mind-sets. And as such, the organisation learns, and changes.

- The significance of organisational culture. **Organisational culture** can be a barrier or change agent in itself. How the organisational culture considers issues such as trial and error learning, reward for learning and change, changes to the status quo or 'the way of doing things around here', the value of human capital and so forth are all essential when considering ODev strategies and programmes.

- Humanistic values. An inherent consideration of **human dignity and development**, as opposed to authoritarian and alienating values.

> ## A humanist approach – common principles in both ODev and Agile Methodologies
>
> Humanistic approaches focus on the self-concept a person has of themselves. A motivated employee is often driven by the nature of the job and the enrichment of the work – the challenge and opportunities for personal and professional growth (see Ramlall[118] for a review on motivation factors). Motivated employees require a manager who guides, facilitates, removes barriers to performance, and offers challenges. This aligns to the Agile method of manager as facilitator rather than director, a focus on constant learning and adapting, and quick feedback to ensure success.

AGILE METHODOLOGIES ARE ALL ABOUT MIND-SET AND CULTURE

Adopting Agile methodologies starts with a change in mind-set. No matter the practices contained in the methodology, if an Agile mind-set is not adopted and included, then the methodology will fail. There are a number of shifts in ways of working and culture that must take place in the move to Agile:

- **Customer-centricity**: the customer must remain at the centre of all that you do. This provides the base and stability on which Agile practices can be built.
- **Mind-set change**: new ways of thinking, learning and collaborating are required. This includes the creation of psychological safety and the ability to make mistakes and learn from them in a non-punitive environment.
- **Leadership buy-in** and championing of an Agile approach: this includes the move from a command-and-control approach to an enabling and facilitative approach.
- **Consistent adoption of Agile practices**: the rigour with which these practices are implemented and embedded will impact success.
- **Requisite skills**: Agile methodologies call for different skills, many of which are scarce. A skills audit needs to be conducted along with strategic workforce planning to ensure an organisation has the right people with the right level of skills.

For OD, there is a need to consider the culture, mind-set and values of the organisation that are in place, as well as those that are required. This is especially important as Agile methodologies introduce not just change, but **constant change**, and change is almost always disruptive. ODev must support the move towards this constant change by instilling flexible, adaptable mind-sets.

There is a vast body of literature on organisational culture and it is easy to get sucked into a highly theoretical analysis which provides more questions than answers. ODev practitioners must consider the current culture within which Agile methodologies are being implemented, and also define the required culture. Identifying the gaps between the current and required culture is a first step in understanding what needs to change and how difficult the change will be. Keep it simple – you do not need complex models to understand what works and what needs to change.

One of the most fundamental shifts is moving from individual contribution to team contribution. Working in small, cross-functional, collaborative teams is key to the success of Agile practices. These teams are not organically formed at first and employees may struggle with the high-touch, intense team environment. Team cohesion, with strong trust, is an essential contributor to team performance. At the same time, the team needs to guard against team think which may work against the introduction of new ideas, alternate ways of thinking, or challenges to the way of working.[120] It is the role of ODev to create team effectiveness interventions to help teams find ways to norm and storm and adapt to Agile practices.

Defining Organisational Culture

Just like people, every organisation has its own unique personality. This unique personality is referred to as the organisation's culture. Organisational culture includes the norms and values of the organisation, as well as the behaviours of employees. It is essentially HOW things are done in an organisation. It also includes the working language, systems, symbols, beliefs and habits of employees.

Shared values have a strong influence on the people in the organisation and dictate how they dress, act, and perform their jobs.[119]

The importance of feedback

Team effectiveness is a significant area of focus for ODev practitioners' Agile transformations. Both the softer skills of communication, collaboration and negotiation, as well as the more technical skills projects demand, are important. The value of feedback cannot be overstated, thus feedback mechanisms should be designed and built into all interventions. The culture journey that ODev will help the organisation navigate will rely on feedback loops not only with customers, but with employees. When soliciting feedback it is important to go beyond asking the same old questions which ask whether the intervention was successful and if the employee feels they have benefitted from it.

Agile practices rely on feedback of HOW employees have learned, not just WHAT they have learned. There is also an expectation that employees share their learning with peers and proactively engage in communities of practice. This makes the 'reach' of the intervention far greater and also ensures that the content of the intervention has been good enough to create its own momentum.

Ask employees if they have shared their learnings with peers. If they have not done so, ask why not. Sometimes asking these questions prompts future action and encourages a learning culture.

Feedback also needs to evaluate effectiveness against individual gains, team gains and organisation gains. If employees are unable to see how their team or the organisation has benefitted from the intervention, the objectives have either not been clearly articulated, or the relevance of the intervention should be questioned. One of the primary responsibilities of ODev is to help connect the dots for employees and make sure they are able to see the 'big picture' and how they both fit in and contribute to organisational success.

How to retain managers in Agile Organisations

Managers often struggle to let go of authority as a tool, thus questions are raised about the necessity of retaining managers in Agile organisations. If the work is conducted by self-managed, autonomous teams, and if hierarchies and silos are broken down, what remains for managers to do? It is essential for managers to understand that their role is not obsolete, but it does need to change.

As Gallup[121] pointed out, there are two main areas where managers need to shift their orientation to their teams:

- **Coordination among teams**: in Agile organisations, teams are fluid, called on for specific tasks, then re-organised. In this situation the manager plays a central coordinating role. Managers must:
  - coordinate team members;
  - remove obstacles and barriers that prevent teams from progressing with their work;
  - help ensure there is no duplication of effort between teams;
  - manage dependencies between teams;
  - enable the sharing of lessons learned across teams;
  - understand team talent and be aware of talent across other areas which they can draw-on for future projects;
  - be aware of opportunities elsewhere in the organisation for team members to take advantage of; and
  - develop team members, giving performance feedback quickly and constructively.
- **Maintaining continuous learning**: the development of employees lies at the heart of Agile. An organisation is considered adaptable, flexible and speedy if that is how its employees behave. To create employees who are quick, adaptable, and open to problem solving, the manager, with the help of HR and the ODev practitioner, needs

to enable continuous learning and development for employees. The emphasis on continuous learning includes the following:

○ Continuous coaching from managers and team leads – it is important to get feedback fast and to make changes quickly.

○ Investment in learning platforms and content which promotes accessible and continuous learning from the organisation. This requires a conscious decision to invest resources (time, money and innovation) in terms of coaching, mentoring and informal and formal training.

The barriers to change

Agile methodologies challenge traditional management styles, hierarchy and positional power. How do leaders manage the loss of power and learn to get their hands dirty again after years of 'managing' and not 'doing' the work? It is a tough change for many to make and there are instances where some leaders and employees simply do not fit in and cannot make the change. Many organisations have tough decisions to make when implementing Agile at scale. Some, like the Dutch Bank ING, recognised that a large proportion of their workforce was not suited to Agile practices and these employees were not retained.

Agile methodologies fundamentally challenge the hierarchy of the organisation. Power is traditionally directed from the top down, and decisions and strategies are typically set by senior management and then 'cascaded down' for implementation. While governance and risk can be reconceptualised and managed in more Agile ways, the role of ODev is to guide the more complex breakdown of the power dynamics and hierarchical relationships when Agile methodologies are adopted. ODev has been criticised in the past for not effectively challenging the status quo, and being unable to manage power dynamics, culture and hierarchical structures.[122] These dynamics constitute significant barriers to Agile transformation and the responsibility of ODev is heightened during this time.

THE LEARNING ORGANISATION

Chandler defined a learning organisation as "one that continuously facilitates learning for its people and transforms as needed. Key characteristics of a learning organization are systems thinking, challenging the status quo, continued growth for teams and individuals, and common understanding of a shared vision".[123]

When an organisation moves beyond an approach that focuses on training for new skills and content to a learning culture, employees are encouraged to:

- refresh their knowledge;
- become skilled in new technologies; and
- enhance capabilities after changes have occurred in their context.[124]

The promotion of a learning culture leads to an organisation with the vision to move fast, provide services quickly and effectively, and be relevant to customers. In addition, this agility and culture of learning should not be seen as the end goal, but rather a 'new way of being' to remain competitive and adaptive.[125]

When developing a learning strategy, the organisation should design for agility. This means that:

- all learning should be embedded into work as far as possible;
- learning should be informal and formal;
- peer learning should be encouraged;
- employees should be given time and access to benefit from continuous learning;
- learning must be easily accessible and 'always-on'; and
- there should be frequent evaluations of learning to ensure the effectiveness of the content and the platforms used.

Chandler[126] argued that formal learning should not be the norm – rather there should be space for on-the-job learning to happen as and when it is needed. There are a variety of ways that this kind of learning can take place, including:

- short videos, online instruction, and mobile apps – anything that allows the employee to access the information they need almost instantly;
- access to a mentor – this could be in the organisation or online; and
- systems and processes that support the above, which are quick and efficient. It is important to set up spaces and times for sharing. Agility includes sharing knowledge and learning among team members and across the organisation.

Other interventions that could also be used include:

- job shadowing, secondments and internships;
- small group tutorials, discussion forums and lunchtime learning sessions (brown bag sessions); and
- communities of practice – online or in teams.

Above all, there needs to be a space where people can fail, as this also generates lessons to learn from. As Chandler noted:[127]

"Think about how often a successful initiative is scrutinized: pretty infrequently. The typical after-event action for a success is a celebration, while failure is typically scrutinized and people held accountable. The problem with both of these situations is what happens to the human psyche. When success is celebrated and failure is punished, a result can be prolonged admittance to failure or hiding of failures altogether, which can lead to greater failures at higher costs."

What does Agile demand of employees?

Gallup[128] recently released a special edition of their *The Real World of Work* publication, which is dedicated to Agile practices. In this edition they define agility from an employee perspective as "employees' capacity to gather and disseminate information about changes in the environment, and respond to that information quickly and expediently". They conducted a survey of employees in the UK, Spain, France and Germany in 2018 on how Agile business is, and the researchers found that participants felt there was much room for improvement.

The researchers asked whether employees felt their organisation had the right mind-set to respond quickly to business needs and whether they had the right tools and processes to respond quickly. They found that between 49% and 60% felt their organisation was not Agile. Importantly, the researchers also found that if employees thought their organisation was Agile, they were also more likely to have confidence in its financial future.

Gallup[129] argued that there are eight characteristics associated with an organisation that is Agile, which are grouped into three main areas:

- Speed and efficiency.
- Freedom to experiment.
- Communication and collaboration.

The following table outlines each of these areas and associates it with the requirements of employees in terms of skills. This starts a process whereby ODev practitioners can consider what skills employees (including team leaders and managers) may require for Agile, which areas require up-skilling, and where these skills may already exist in the organisation.

Table 5.1: Skills and behaviours for an Agile workplace

Agile Characteristic	Skills and Behaviour
Speed and efficiency	• Speed is considered a sign that an employee can problem solve and is empowered to do so (and is most likely trusted to do so). • Speed is associated with employee empowerment, decentralised decision making and a focus on procedural simplicity.
Freedom to experiment	• Creativity and ingenuity are needed. This requires the trust and space to be able to do this. • There cannot be a fear of failure in this setting, or a fear of speaking out.
Communication and collaboration	• Make sharing knowledge and experience routine. • Cross team coordination and communication is essential to make sure processes and experiments are not repeated. • Of importance here is that the information does not just flow from the usual hierarchy or manager. Informal and formal networks are part of the communication and collaboration network. Managers as the facilitators need to understand these roles and where they exist, and tap into them.

A recent study by Omar et al.[130] investigated the skills required to succeed in an Agile Scrum Team (Scrum is a commonly used Agile Methodology). It was found that a distinction between hard and soft skills was required (see Table 5.2).

Table 5.2: Hard and soft skills required in an Agile Scrum[131]

Hard Skills	Soft Skills
Programming language	Analytical skills
Written and spoken language	Communication skills
Database	Interpersonal skills
Expert area	Leadership skills
Scrum role experience	Teamwork skills
Scrum hours	Thinking skills
Number of sprints	
Scrum knowledge	

An example of how these hard and soft skills are translated into action is illustrated in Figure 5.2. below.[132]

General Information			
Employee Name	Liza Khasasi	**Employee ID**	1

Hard/Technical Skills

Criteria	Detail	Proficiency
Programming Language(s)	HTML	Beginner
Language(s)	• Behasa Malaysia • English	• Intermediary • Advanced
Database(s)	MS Access	Beginner
Expert Area(s)	• MS Office – Word • MS Office – Excel • MS Office – PowerPoint • MS Office – Visio • MS Office – Project	• Beginner • Beginner • Beginner • Beginner • Beginner
Scrum Role(s)	Scrum Master	
Scrum Hour(s)	100	
Number of Sprint	3	
Scrim Knowledge	Advanced	

Soft Skills

Criteria	Detail	Proficiency
Analytical Skills	The ability to visualize, articulate and solve both complex and uncomplicated problems and concepts and make decisions that are sensible and based on available information.	Intermediary
Communication Skills	The ability to convey information to people clearly and simply in a way that means things are understood and get done. It is about transmitting and receiving messages, and being able to read audience.	Intermediary
Facilitation Skills	The ability to guide group member in meeting to share ideas, opinions, experiences and expertise in order to achieve common goal and agreeable action plan.	Intermediary
Interpersonal Skills	Ability to practice life skills for everyday communication and interaction with other people (individually and in groups) in both professional and personal lives.	Beginner
Leadership Skills	Ability to influence, aid and support others to accomplish an objective.	Beginner

Criteria	Detail	Proficiency
Management Skills	Ability to make business decisions and lead subordinates (plan, organize, direct and control) within a company.	Beginner
People Skills	Ability to use both psychological skills and social skills to communicate effectively with people in a friendly way, especially in business.	Beginner
Planning Skills	Ability to look ahead and accomplish goals or avoid emotional, financial, physical or social hardships to make and implement decisions.	Beginner
Teamwork Skills	Ability to work in team and perform by combining individual talents (skills) with others to accomplish goals.	Beginner
Thinking Skills	Ability to employ mental processes to do things like solve problems, make decisions, ask questions, make plans, pass judgements, organize information and create new ideas.	Beginner

Figure 5.1: Example Articulation of Hard and Soft Skills Required for a Scrum Team

It is important to note that each skill is rated on a proficiency scale from beginner to expert. As organisations reduce hierarchy, proficiency is emerging as a way in which career progression can be managed without vertical, linear career paths. In an Agile organisation, the way to increase your pay and position is to improve your proficiency. This drives very different behaviour to traditional career paths, which often require extensive stakeholder engagement, the management of larger teams, and political manoeuvring to get to the top.

The decision to implement Agile methodologies and become a learning organisation seems like a positive proposition that will benefit employees and organisations. In practice, however, there are many employees who find this change threatening and anxiety-provoking. Conboy et al.[133] outlined several ways in which employees may feel under-equipped and 'shown up' as their organisation moves towards implementing Agile. In a qualitative investigation of large organisations which moved to Agile – both successfully and with some challenges – the authors found that some of the people challenges included the following:

- Some people felt that the Agile environment showed up their **skill deficiencies**. For ODev practitioners this suggests that learning and development needs to include change management programmes to point out that this skill deficiency may emerge. There should be appropriate up-skilling, coaching and mentoring interventions available should employees so require. Managers should specifically

71

look for when employees feel 'shown up', as this may result in lowered self-esteem and withdrawal, impacting performance.

- A perception that team members should be a **master of all trades**. Instead of specialists and skill sets that have clear boundaries, Conboy et al.[134] found that some people felt that they needed to be cross-functional skill specialists. Managers found it very difficult to find this multi-skilled talent in the organisation. The authors noted that some organisations responded by sending employees on a number of skills courses, whereas before these would have been divided among team members – an expensive and unsustainable solution. The authors suggest a balance between being a 'master of all trades' and being a specialist. Moving across Agile projects will lead to cross-functional skills development, but it is unreasonable to expect all employees to know all things.

- A greater **emphasis on social skills** was found. This study found that Agile demands more face-to-face communication, more feedback to groups, and greater engagement and debate. In certain environments social interaction may not be a key strength. In these instances, ODev practitioners should consider facilitation and presentation training, as well as the development of communication and negotiation skills. The authors cautioned, however, that it may well be that certain individuals will not take to social interactions, and a different approach needs to be found or productivity will suffer.

CONCLUSION

Organisation development plays a key role in driving and enabling an Agile culture, yet the change to an Agile mind-set is not one that comes naturally to many. Although culture is an outcome, without the right inputs, one cannot direct the culture journey. The role of ODev is to challenge traditional hierarchies, help break down silos, promote effective teams and manage the change so that employees and leaders understand what is required of them. Helping leaders to function from a position of influence and not positional power is a key challenge for ODev.

Agile methodologies flourish in a learning organisation, but one should not under-estimate the challenges attached to this – balancing development opportunities against productivity; the cost of learning against effectiveness in implementation; the people change required if Agile is poorly implemented and people become disengaged or frustrated; and the productivity impact of mistakes if these are large and costly. The ODev practitioner's role is to balance these concerns, ensure the method of intervention is correct, understand the current versus desired culture, and provide feedback to Agile leaders and executives on how to navigate the culture journey.

CHAPTER 6

MEASURING AND DRIVING TEAM PERFORMANCE

INTRODUCTION

Today more than 70% of all employees work in service or knowledge-related jobs. Their performance is driven by their skills, attitude, customer empathy, and by their ability to innovate and drive change by working through teams. These skills must be built over time, and successful performance management must be focused on constantly developing these capabilities rather than ranking them at a moment in time.[135] There have been significant changes in traditional performance management approaches over the last decade, but the advent of Agile methodologies calls for even greater change. The complexity of promoting, managing and appraising team performance is the key focus of this chapter.

THE TRADITIONAL APPROACH TO PERFORMANCE MANAGEMENT

Performance management is a process that measures employee performance against set performance standards. The objectives of performance management are defined as follows:

- To align organisational and individual goals.
- To foster organisation-wide commitment to a performance-oriented culture.
- To develop and manage the human resources needed to achieve organisational results.
- To identify and address performance inefficiencies.

- To create a culture of accountability and a focus on customer service.
- To link rewards to performance.[136]

The traditional performance management process is a cascade of the strategic and operational requirements of the organisation down to an individual level. Each individual needs to understand fully and exactly what and where he or she contributes to the final result in the total picture for the organisation.

The WorldatWork approach describes the traditional performance management process as follows:

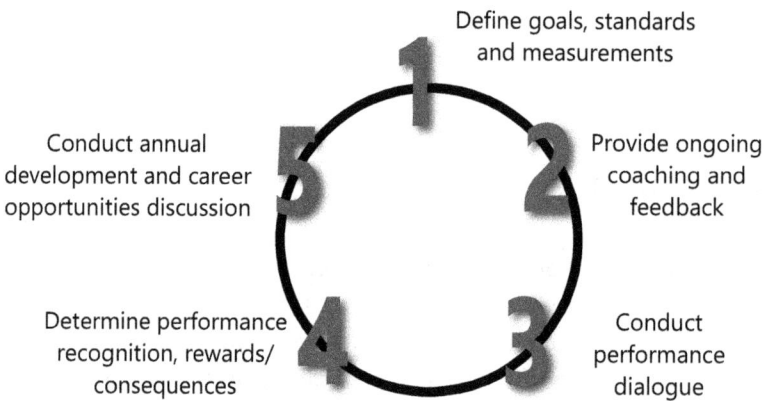

Define goals, standards and measurements

Conduct annual development and career opportunities discussion

Provide ongoing coaching and feedback

Determine performance recognition, rewards/ consequences

Conduct performance dialogue

Figure 6.1: Performance management process phases[137]

The performance management process is designed to involve both managers and reports in each part of the cycle. The cycle can take place quarterly, semi-annually or annually, however the most common approach is twice a year. Notice that the process is a circle; it is an unbroken cycle of communication between the manager and employee.

The phases of the performance management process are as follows:

Phase 1: Define goals, standards and measurements

Phase 2: Provide ongoing coaching and feedback

Phase 3: Conduct performance dialogue

Phase 4: Determine performance recognition, rewards or consequences

Phase 5: Conduct annual development and career opportunities discussions

Typically, phases 1 through 4 cover the actual process that is usually addressed within the total rewards spectrum of an organisation. One of the most important aspects of the performance management process is actually phase 5 – the individual development and career opportunity offered by the process. Far too often, the cycle is linked to a reward-management process rather than a development process, and so necessary synergies are lost. Often employees will "tune out" of the performance discussion, waiting for the rating and what that means for their remuneration, bonus or position.

Much of the recent literature actually stresses holding the above phases at different times, but also emphasises coaching and feedback. Practically, this would mean that phase 2 on coaching and feedback is given a lot more prominence, even within the performance discussion. To ensure consistent improvement and development, the feedback – both positive and negative – must be as close to the event or behaviour as possible, and must be constructive.

The performance dialogue can then focus on the performance that is linked to the ultimate achievement of the goals. As the continuous feedback has been happening throughout the performance period, there should be no surprises for the manager or employee. It is also recommended that the performance discussion happens at a different time to the discussion about remuneration – even a month apart. If the two are done together, employees often stop listening to the review and focus on the money discussion, and the process loses some of its effectiveness, leading to frustration.[138]

TEAM PERFORMANCE MANAGEMENT IN AGILE SETTINGS

In Agile methodologies, the unit of production moves from the individual to the team. In order to promote collaboration and ensure that teams understand and invest in team performance, we need to look at a new and bespoke form of performance management. From goal setting to providing continuous feedback, appraisals and rewards, the entire performance management process has to become more flexible and also align with Agile principles.

The cornerstones of Agile performance management are illustrated in the following shifts:

Table 6.1: Traditional vs. Agile performance management

Traditional Performance Management	Agile Performance Management
From	**To**
Individual	Team
Annual goal setting	Flexible goal setting
Annual/Biannual review	Continuous review
Backward looking	Forward looking
Uses management practices	Uses coaching practices
Often includes feedback from line manager only	Includes 360-degree feedback
Primary purpose is evaluation	Primary purpose is development

Let's have a look at these shifts in more detail.

From an individual focus to a team focus

In Agile teams, roles are blurred and goals are assigned to the team and not the individual. When looking at performance, the team either succeeds or fails as a unit. It does not matter if any one team member is a star performer or poor performer, the team is evaluated as a whole. When looking at the measurement of team performance, there are two clear areas which must be addressed:

Team results

The team is measured on its work results or output. These types of measures could include the delivery of a working solution; the use and acceptance of the team's final report/solution; as well as the ability of the team to deliver results within time and budget constraints.

Team processes

The team can be measured on its internal group dynamics. These types of measures could address how well the team works together as a group; the effectiveness of team meetings; the ability of the team to reach consensus; and the team's problem-solving techniques.

A combination of both team results and processes is necessary to evaluate the overall performance of a team. In Agile, how you produce results is as important as the results themselves.

From annual goal setting to flexible goal setting

Goals have always been a key element of performance management, but traditional practices were rigid and goals were set annually. Annual goals do not feature in Agile methodologies. One of the primary purposes of Agile is to break up the work into smaller bite-sized chunks and focus on the completion of each piece of work iteratively. There is an acknowledgement that learning and development take place continuously and that customer feedback loops lead to changing requirements. It is thus necessary to have flexible goals that can be adapted as and when required. The changing of a goal should not be seen as a failure to achieve the original goal, but rather a success in identifying the necessity to make the changes required.

Goals should help employees manage their weekly, monthly, and quarterly objectives. While goals can be changed when required in Agile practices, most teams set goals quarterly. Research shows that companies that manage goals quarterly generate 30% higher returns from that process than companies that manage them annually.[139] The process of setting up goals should be a mix of top-bottom and bottom-top approaches, making the exercise truly collaborative.[140]

From annual/biannual reviews to continuous reviews

Traditional performance management systems most commonly have two formal reviews/ appraisals per year. This leads to issues with both the validity of the process as well as the spirit in which the appraisal is conducted. The primary problem with validity is the recency effect, where managers are most likely to let recent events influence their perception of the entire six month review period. Reviews are often dreaded by employees and are approached with anxiety. Both of these factors reduce the effectiveness of the reviews.

In Agile methodologies, feedback is provided on a continuous basis. Agile ceremonies/ practices are used to facilitate this feedback regularly. Daily stand-ups (short meetings where progress is tracked and daily activities are planned) provide an opportunity to assess delivery against goals and time lines, and the use of artefacts like Kanban Boards make delivery, or the lack thereof, transparent.

An Agile sprint is a regular, repeatable work cycle during which work is completed and made ready for review. Sprints periods can vary from 2 weeks (most common) to 30 days.[142] Sprint retrospectives are sessions which take place directly after each sprint. In these sessions, the team will review their performance during the last sprint and identify actions for improvement going forward. Retrospectives create an ideal platform for feedback to be shared by team members.

Most Agile methodologies work in three month cycles. At the end of each cycle there is a Quarterly Business Review, where the work of the team is evaluated by a panel of subject matter experts and leaders.

> **Kanban Boards**
>
> "Kanban" is the Japanese word for "visual signal". A Kanban board is a tool designed to help visualise work, limit work-in-progress, and maximise efficiency (or flow). Kanban boards can be physical or digital, using cards, columns, and continuous improvement to help teams commit to the right amount of work, and get it done.[141]
>
>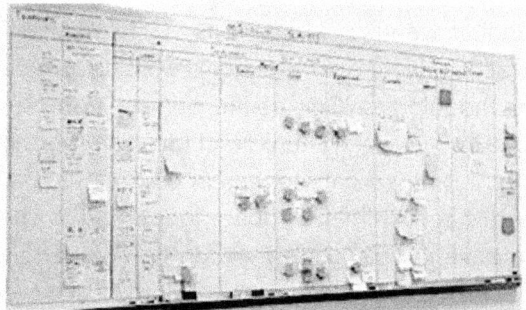
>
> Column headings can be as simple as (Backlog work to be done), In Progress, Testing, and Done.

During these review sessions, teams are assessed on whether their goals were met, the quality of their results, the commerciality of their product, and whether timelines and budgets were met. This forms an additional opportunity to provide feedback to teams.

Agile practices create multiple platforms for continuous feedback, which form the basis for continuous reviews. Poor performance is dealt with as it arises and development or sanctions can be instituted immediately, as opposed to having to wait for a formal annual/biannual performance review.

From backward looking to forward looking

When performance feedback is included in Agile practices, it is generally discussed in terms of how to adjust delivery to better accomplish the activities for the next day/sprint/quarter. The focus is thus forward looking as compared to traditional performance feedback, which tends to look back and focus on past work/successes/challenges.

From using management practices to using coaching practices

It is not only the frequency of feedback that increases in Agile methodologies, but also the nature of the feedback. Instead of using traditional management practices, Agile practices promote the use of coaching. Management practices and conversations tend to focus on current issues or problems in the workplace. The onus is placed on the manager to resolve problems and provide directive feedback on how the employee can improve.

Solution-focused coaching restores ownership to the employee by ensuring that every discussion reinforces the desired result. This requires ongoing and open communication, preferably using regular, face-to-face conversations. Coaching cultivates a problem-solving competency in employees and positively influences energy and morale.[143]

From line manager feedback to 360-degree feedback

360-degree feedback, otherwise known as multi-source or multi-rater feedback, is a comprehensive and structured way to obtain feedback. It is a system or process in which employees receive confidential/anonymous feedback from the people who work around them. This typically includes the employee's manager, peers, direct reports, and, in some cases, customers or clients – in fact, anybody who is credible to the individual and is familiar with their work can be included in the feedback process. This is usually in addition to completing a self-assessment on performance. Each source can provide a different perspective on the individual's skills, attributes and other job-relevant characteristics, and thus help to build up a richer, more complete and accurate picture than could be obtained from any one source.[144]

A 360-degree feedback tool assesses the contextual component of performance, that is, it assesses **how** employees go about completing the tasks they do in their jobs, rather than whether or not these tasks were completed. 360-degree feedback can also be a useful development tool and help employees get a better understanding of their strengths and weaknesses. Similarly, employees can give regular feedback to their managers. Highly effective managers accept and welcome feedback and use this feedback to improve their management and leadership competencies.

In Agile methodologies, 360-degree feedback is best provided continuously and not via a scheduled survey, as is common when 360-degree feedback is used as part of a traditional performance management approach. To enable continuous 360-degree feedback, it is recommended that organisations invest in a digital system. This allows employees to capture real-time feedback as close to the event or time as possible in an easy and recordable format. There are a number of mobile apps that now cater for such feedback, promoting ease-of-use and convenience.

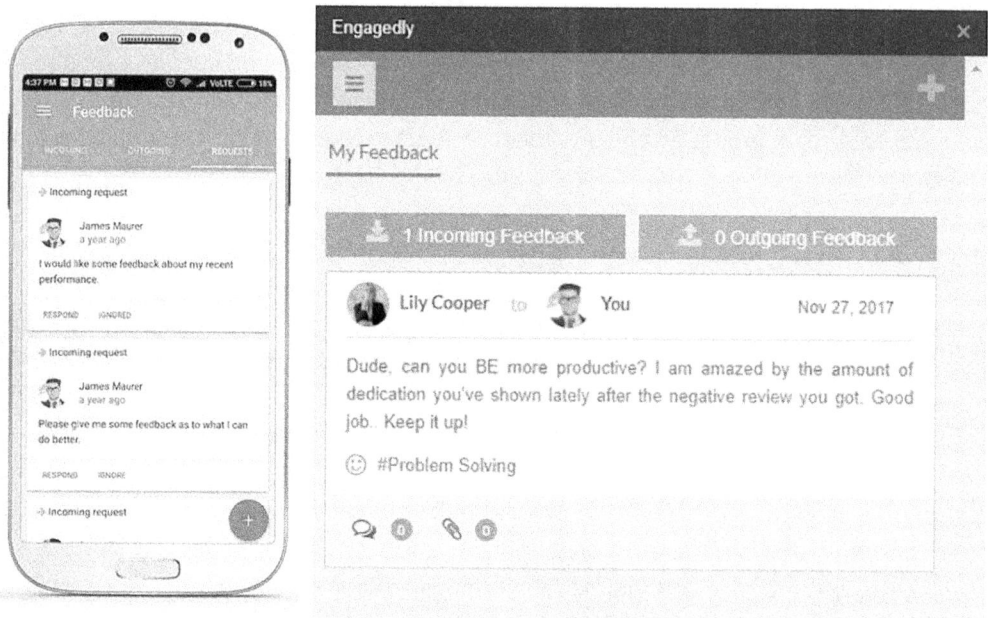

Figure 6.2. Digital 360-degree feedback

Digital 360-degree feedback tools, like the one pictured above from Engagedly, make it easy to ask for feedback and send feedback in real time.[145]

From a primary purpose of evaluation to a primary purpose of development

Another difference between traditional and Agile performance management is the ultimate goal of each method. Traditional performance management is used primarily to evaluate performance with the aim of assigning a score/rating that can be linked to reward. The appraisal seeks to justify the score/rating by exploring an employee's weaknesses and where they fell short. While intended as constructive criticism, the experience is often perceived by employees as punitive and negative.

Agile performance management prioritises growth and development, where successes are celebrated and losses are used as teachable moments in real-time. The fact that feedback is provided on a regular basis facilitates on-the-job learning. The main purpose of Agile performance management is to enable successful performance through ongoing development.

INDIVIDUAL PERFORMANCE MANAGEMENT IN AGILE SETTINGS

Team performance management does not replace individual performance management in Agile settings, but adds to it. It is rare to find an Agile team where team members do not have individual line managers. This is important for a number of reasons:

- Agile teams are cross functional and team members require a subject matter expert to provide content/domain guidance. In some instances, this role is played by the line manager; in others there are 'chapter' or 'domain' leads that fulfil this role.
- In some Agile models, employees are placed in resource pools and are then assigned to teams as business demands are identified and formalised. Resource pool managers can play the role of a line manager, getting to know all members of their resource pool well enough to know their strengths and being able to allocate them to teams as required.
- Some team members may not be allocated 100% to the Agile team and may have other goals that must be managed.
- Team members still need a line manager for administrative proposes to approve leave and manage other human resource tasks.

In Agile performance management, regular touch points with the line manager constitute an essential component of the coaching and continuous feedback cycle.

A MODEL FOR AGILE PERFORMANCE MANAGEMENT

Aligning the Performance Management cycle to the Agile Business Development cycle provides a working model of how one can implement an Agile performance management model.

Table 6.2: A model for Agile performance management

Time Period	Beginning of the quarter	Daily	Every 2 Weeks	End of the Quarter	End of the quarter	
Team Activities	Team Goal Setting	Continuous Feedback at stand-ups	Continuous Feedback at sprint retrospectives	Quarterly Team Performance Review	Quarterly Business Review	Process repeats each quarter
Individual Activities	Individual Goal Setting with Line Manager	Continuous feedback via real-time 360-feedback and monthly check-in conversations with line manager		Quarterly Performance Review with Line Manager		

81

Team activities

Team goal setting: at the beginning of each quarter, the team meets to discuss and agree their quarterly goals. This is usually linked to a defined project or a specific component of a project. The team agrees on ways of working and measures of success for their delivery. This could include time, budget, and output. Goal setting is not done in isolation and is usually guided by previously agreed parameters set in the preceding quarterly business review. Goals are set quarterly but remain flexible. If customer requirements change or new challenges and dependencies are discovered before the end of the quarter, the team can pivot and adapt their goals as required.

Continuous feedback at daily stand-ups: team members meet daily for short meetings to keep track of progress, decide on daily activities and drive alignment. These stand-ups provide an ideal platform for teams to provide feedback in an open and transparent manner with the aim of meeting goals and motivating accountability. The stand-ups also provide a place where team members can ask for support if they are struggling and work needs to be rebalanced or development/expert skills sought.

Stand-Ups

The daily reinforcement of sharing individual successes and plans keeps everyone aligned and excited about the team's overall contribution and progress.[146]

A stand-up is a daily meeting that involves the core team. Stand-ups should be limited to 15 minutes and are conducted at the same time every day. These three simple questions can be used to generate structure:

- What did I work on yesterday?
- What am I working on today?
- What issues are blocking me?

(Photo source: Agile Connection)[147]

Continuous feedback at sprint retrospectives: at the end of each sprint (which optimally lasts for two weeks) the team meets to discuss the past sprint, what was successful, and what could have been done better. This discussion should include ways of work and facilitate feedback between team members and other resources who have supported or worked with the team. It can also be useful for line managers to be invited to these retrospectives so they can stay close to the work and team dynamics.

Quarterly team performance reviews: once a quarter, the team should meet with the express purpose of conducting a team performance review. At this review the team should decide on whether they have met their quarterly goals and if so, whether they have exceeded expectations. Organisations use different ratings when assessing performance – some use numerical scores while others use qualitative descriptors. The rating must include an assessment of WHAT the team has achieved as well as HOW the team has worked together. The weighting between the two can vary but most organisations weight the output more heavily. The assumption is that if the team did not work well together, it is unlikely they would have met their goals as successfully, thus the HOW is reflected in the output and not just as a stand-alone indicator. This session serves as a team self-evaluation and together the team decides on a rating, which is applied to all team members regardless of their individual performance.

Quarterly business reviews (QBR): the QBR takes place at the end of each quarter, when each team presents their output to a panel of leaders and subject matter experts. They are then rated on whether they have met their goals and whether they have delivered within their time and budget constraints. Each team has an opportunity to pitch for continued funding and work if their product is not complete or needs a further iteration. At the end of the QBR, the panel decides which teams remain together and which are dissolved, and they also agree on new deliverables, budgets and timelines.

Individual activities

Individual goal setting with line manager: individual employees meet with their line managers at the beginning of each quarter to agree their goals for the quarter. This discussion will include a personal development planning session, where employees have an opportunity to request formal development and discuss career aspirations. The line manager also uses this opportunity to conduct a talent review, establish retention drivers and update succession plans for roles characterised by scarce and critical skills. If employees are dedicated to the Agile team, the line manager can consider adding goals that talk to the sharing of skills and mentorship of other employees. If the employee is not dedicated to the Agile team, other deliverables are articulated as separate goals. Despite the fact that this goal setting process is conducted quarterly, goals should be flexible and open to change based on mutual agreement between the employee and the line manager.

Continuous feedback via real-time 360-degree feedback and monthly check-in conversations with line manager: formal and informal feedback is encouraged throughout the quarter, as it is well-established that feedback (both positive and constructive) is best provided as close to the event as possible. Real-time feedback should therefore become an established way of working. This regular 360-degree feedback serves to improve the quality of the work and ways of work, and also helps to create an environment of psychological safety where employees are able to challenge the status quo, ask for help when necessary, and feel unafraid to admit failure. The 360-degree feedback should be supplemented with monthly check-in conversations with line managers. These conversations do not have to be formal but help establish a rhythm to embed the practice of continuous feedback. These conversations become even more important where line managers are removed from the team work and are not close enough to the detail.

Quarterly performance review with line manager: once a quarter, every employee should have a formal performance review with their line manager. Employees should complete an individual self-evaluation in preparation for this session and take this along with their team self-evaluation and QBR ratings. Together with their line manager, they will then agree on an appropriate rating for the quarter. Since the formal review will have been preceded by numerous check-in conversations, the sessions are likely to be short with no surprises. It is important to have these quarterly sessions as Agile teams are not always constant and employees may be moved to different teams and/or different projects each quarter based on business demand.

The proposed model of Agile performance management takes team and individual performance into account and is characterised by ongoing feedback and flexibility.

(CASE STUDY) **Using continuous feedback to drive performance management**

Adobe is an American multinational computer software company that has historically focused on the creation of multi-media and creativity software products, with a more recent expansion into rich Internet application software development. It is best known for Photoshop, an image editing software; Acrobat Reader; the Portable Document Format (PDF); Adobe Creative Suite; as well as its successor, Adobe Creative Cloud.

Adobe was one of the first organisations to abandon annual performance appraisals in 2012. They described their old process as so complex, bureaucratic, and paperwork-heavy that it ate up thousands of hours of managers' time. It also created barriers to teamwork and innovation. Employees often felt undervalued and uninspired as their performance ratings were force-ranked and used for compensation.

Adobe replaced their old performance reviews with regular check-ins, supported by frequent feedback – both positive and constructive. There are no performance ratings or rankings and they allow different parts of the organisation to determine how frequently they should hold check-in conversations, according to their work cycles.

The table below illustrates the differences between Adobe's old and new performance management processes:

Table 6.3: Adobe's old vs. new performance management processes

	Before: The annual performance review	**After: Check-in**
Setting priorities	Employee priorities set at the start of the year and often not revisited.	Priorities discussed and adjusted with manager regularly.
Feedback process	Long process of submitting accomplishments, soliciting feedback, and writing reviews.	Ongoing process of feedback and dialogue with no formal written review or documentation.
Compensation decisions	Onerous process of rating and ranking each employee to determine salary increase and equity.	No formal rating or ranking; manager determines salary and equity annually based on performance.
Cadence of meetings	Feedback sessions inconsistent and not monitored. Spike in employee productivity at the end of the year, timed with performance review discussions.	Feedback conversations expected quarterly, with ongoing feedback becoming the norm. Consistent employee productivity based on ongoing discussions and feedback throughout the year.
HR team role	HR team managed paperwork and processes to ensure all steps were completed.	HR team equips employees and managers to have constructive conversations.
Training and resources	Manager coaching and resources came from HR partners who could not always reach everyone.	A centralised Employee Resource Centre provides help and answers whenever needed.

For Adobe, a good check-in includes three elements of discussion: expectations, feedback, and growth and development. When each of these areas have been discussed, managers and subordinates know they have had a meaningful conversation.

Expectations: refers to the setting, tracking, and reviewing of clear objectives. Both employees and line managers agree on roles and responsibilities for the objective, and are aligned in how success will be defined.

Feedback: refers to ongoing, reciprocal coaching on a regular basis. Feedback is the logical next step from a discussion about expectations. The reciprocal nature of feedback is emphasised. Feedback conversations provide answers to two questions: 1) What does this person do well that makes them effective? and 2) What is one thing, looking forward, they could change or do more of that would make them more effective?

> **Growth and development:** refers to the growth in knowledge, skills and abilities that would help employees perform better in their current roles, as well as making sure that managers have understood each of their employees' long-term goals or career growth ambitions, and worked to align those goals with current objectives and opportunities.
>
> A vital part of making check-ins successful is not just their forward-looking nature, but also their frequency. If you are checking in regularly, then it is much easier for both managers and employees to see progress.
>
> The result has been a marked increase in employee engagement, with voluntary turnover decreasing by 30% since check-ins were introduced.[148, 149, 150]

AGILE PERFORMANCE MANAGEMENT AND THE EMPLOYEE PSYCHOLOGICAL CONTRACT

The majority of employees today have had individual performance management serve as a stable component of their working lives. Their psychological contract with their organisation is based on the principle of individualism; they believe that if they work hard and achieve individual success, they will be duly rewarded. Agile methodologies challenge this psychological contract which can lead to fear, mistrust and anxiety. We all have memories of working in groups on school projects where someone did not contribute equally, let the team down, and still received the same mark/grade as everyone else. The fear that this will be replicated in the workplace persists. No-one wants to be disempowered to be able to perform, nor are they happy to carry the weight and work of a poor or disinterested colleague.

Team performance relies on trust, inclusion, diversity, and the blurring of roles to achieve success in Agile practices. While it is true that people have different optimal styles of working, resistance to working in teams is not an adaptive response to the new world of work. A willingness to adopt new ways of work and a readiness to experiment are prerequisites for success in Agile methodologies. There may be pockets in organisations where employees who prefer not to collaborate or work in teams remain, but over time, these will not be the roles of the future or the roles that will lead to career advancement and success.

Changing to a model of team performance can be difficult, but all employees are strongly encouraged to let go of preconceived notions that teamwork is not a part of their psychological contract with their organisation. As the world of work changes, teamwork is likely to become the predominant and most successful way of work.

PITFALLS OF AGILE TEAM PERFORMANCE

Agile performance management allows the team to identify poor performers early on. The theory is that these poor performers will receive extra development to assist them. What happens though, when there are not enough skilled resources available to fill all roles with proficient employees? Development can take years to build proficiency and in the interim, the organisation has to make some tough decisions. Do they build, buy or borrow the skills required? Buying and borrowing skills is expensive and even if there are funds available, there is often a shortage of talent in the market. In these cases, team performance can be jeopardised and the pace of Agile delivery compromised. Agile teams are lean and carrying poor performers can affect productivity and morale. When poor performers are identified, action needs to be taken as quickly as possible.

Another pitfall to be wary of is assuming that velocity is equal across teams; not every team can produce results at the same pace. There will likely be star teams and assuming other teams will have the same pace is going to create problems.[151] Agile practices are still relatively new to most workplaces and as transformation journeys take shape, organisations will want to create internal benchmarks to assess productivity. Performance at the expense of wellbeing is not sustainable and a conscious effort must be made to allow teams sufficient time to recharge. Teams thus need to set their own pace of delivery.

As teams start to successfully adopt Agile practices and see positive results, they may decide to add stretch goals to their sprints. This can become a pitfall when teams either end up working large amounts of overtime to meet the stretch goals, or are unable to meet the goals. Consistently missing stretch goals can be demoralising to the team while having no bearing on the project. Stretch goals can also be misconstrued as they filter up the hierarchy and turn into hard requirements.[152]

The team decides on how much work it will do in a sprint and pressures to over-commit from the team themselves or from external parties can negatively affect performance. The result of such pressure is usually low-quality work and negative team dynamics.

CONCLUSION

As Agile practices become more mainstream in the new world of work, human resource practices will need to evolve to enable these new ways. Managing performance in a developmental and continuous fashion is key to Agile. Employees who work in an environment that promotes flexibility and autonomy are far more likely to be productive, increase their discretionary effort and be receptive to culture change. There have been significant shifts in the field of performance management over the last decade, and Agile practices are simply pushing these changes to the next level.

CHAPTER 7

REWARDING TEAMS IN AN AGILE ORGANISATION

INTRODUCTION

To be a top-performing organisation, an employer must attract and retain employees with a talent management strategy that is clearly aligned to the business' objectives. Reward can be used in every part of a talent management strategy, from attracting individuals to an organisation, through to employee retention and development. It is also used to reinforce the organisational values that are important to an employer, and it reflects the organisation's vision and goals. Performance in Agile organisations is driven by teams and reward practices need to be re-engineered to enable business agility and ensure that team members are not pitted against each other, causing internal competition. The Agile values of transparency, empowerment and rapid feedback are key to developing more Agile reward philosophies and practices, but as you will see in this chapter, finding the right balance between stability and agility in reward can be difficult.

THE CHALLENGES OF TEAM REWARD

In many high performing organisations, pay-for-performance and the notion of meritocracy have become entrenched. We have encouraged, recognised and rewarded individual performance for decades and moving to a team approach will require innovative thinking and concerted change management. High performers have come to enjoy the idea that the harder they work and the more they succeed, the more they are able to earn. The adoption of Agile practices does not mean that individual performance is no longer important; rather we have to figure out how to reward individual performance without encouraging internal competition, local optimisation, or one person feeling rewarded while another feels punished. Compensation should be used to motivate people, but can easily be divisive and have a negative impact on performance.

One of the most significant challenges in team reward is allowing for the variance in personal commitment and expectations of team members. Some employees are happy to come to work, do a good job, and go home. Others want to go the extra mile and really make a difference. Rewarding a team for their collective performance negates the difference in the discretionary effort of individuals. The danger of social loafing is real in many teams. Social loafing is a phenomenon where certain team members get by thanks to the efforts of their colleagues. This can be a huge cost to an organisation, and lead to resentment between colleagues. The consequence for underperforming is left to the team to manage. In addition, where there is social loafing, individual initiative can be undermined and discouraged – there is less motivation to work hard when your team mates do not, but everyone is rewarded equally.

Another common challenge in implementing team reward is the fostering of unhealthy internal competition between teams. Instead of fostering cooperation, the consequences include lack of information sharing and collegiality. Where there is low trust and transparency in an organisation, team reward can be destructive.

It must also be acknowledged that not all teams are created equal:

- Teams have varying degrees of teamwork.
- Some teams operate interdependently, while others are really a collection of individuals who independently carry out similar tasks. This distinction is important when looking at incentive structures.
- Only in truly interdependent teams does a team-based incentive structure work to motivate team members.
- If teams do not need to work interdependently with other teams, a team-based structure often falls flat.

DESIGNING A TEAM REMUNERATION SYSTEM

Create reward principles

The first step to designing a team remuneration system is to agree on a set of reward principles that will guide the approach. The following principles are suggested to help enable Agile practices:

- The system should be **inclusive** – all employees should have access to the reward system. There cannot be some named subset of people able to participate in the plan.
- A mind-set of **abundance** should be adopted – everyone should be able to be at the top every time. Build a system that assumes everyone has the ability to be a top performer.
- **Transparency** is key – whatever you choose to measure should be clearly articulated, visible and tracked so team members can see where they are relative to their peers at all times. The measures and goals have to be clear and attainable.
- **Cooperation** should be rewarded – there cannot be any penalty for working together to achieve a goal. If two people work on a goal together, the reward has to be as much or greater than working on it individually.
- Reward for **sustainable** growth – do not create team incentives that encourage people to only look for short term economic outcomes. This will not help the organisation grow.

A step-by-step plan to rewarding teams

Getting the most from your teams depends on rewarding and recognising them collectively. Most management systems focus on individual performance, undermining the desired team performance. You do not have to overhaul your organisation's evaluation process or pay structure – you can support the right behaviours with the things that are in your control.

Amy Gallo[153] provided a six-step plan to implement team rewards that is intuitive and easy to follow.

Step 1 – Set clear objectives

Team members have to understand and agree on what success looks like. Before the team starts working on a project/product, they need to agree on common goals/ objectives and metrics. A good way to articulate a shared understanding of success is to ask the team to answer the question: What would it take for us to give ourselves an A? This dialogue lays the groundwork for collaboration in an objective way.

Step 2 – Check in on progress

Once the team has started working, check in regularly. These check-ins can be formal or informal, face-to-face, or even allow for anonymous feedback. The best way to elicit feedback is to pose questions to the team to enable them to assess their progress.

These could include: How are we performing as a team? What obstacles can we remove? Another tactic is to ask the team to give themselves a grade to track progress. Not only can you see if the team is aligned in terms of their self-review, but this also allows you to problem solve if the grade is poor and celebrate if the grade is high.

Step 3 – Use the full arsenal of rewards

If there is a way to influence or change the reward structure in your organisation, including a formal component of discretionary compensation for team performance is optimal. If, however, you are unable to make changes at this level, consider the wide range of non-monetary rewards that are at your disposal. You can be creative here and look beyond the traditional team dinners and spa vouchers. Most high performers are incentivised by rewards that enable them to gain experience, learn new skills and get exposure to senior leaders. Consider conference attendance, training/development seminars, job rotation, inclusion in innovative projects, and time with senior leaders as options.[154]

> **Small rewards are often greatly valued**
>
> Mike Cohn described working with a few product owners who carried five-dollar bills and gave them to team members who could quote the project's elevator statement or three main goals when asked. No team member's life was improved by $5. It was more the knowledge that they passed the test when asked. More than one of these team members pinned the $5 to their cubicle wall.
>
> Similarly, a handwritten note can do wonders in this era of constant email deluge. Take the time to write a note every now and then thanking a team member for something special and specific they did.[154]

Step 4 – Get to know your team

Rewards are only motivating if they are valued by the team. Additional learning and development opportunities may be highly sought after by some team members, whilst others may just see this as more work. You need to get to know your team, and team members need to get to know each other. Not only does this enable team effectiveness and cohesion, it also allows you to find ways to incentivise and reward the teams that work.

Step 5 – Focus discussions on collective efforts

When talking to or about your team, focus on their collective efforts and identity. For team rewards to be successful, team members need to see themselves as a part of a unit and not a collection of individuals. This applies to both team successes and set-backs. Talk of individual contributions should be minimised to ensure that the team does not start identifying top and bottom performers. This can be divisive and lead to the rejection of equal team rewards.

Step 6 – Evaluate team performance

Although individual performance reviews will likely remain, you should also introduce a team performance review. This should include both a self-review and a moderated review. The most important thing to remember in the team review is that it is "one for all and all for one". The team either meets its goals as a unit or it does not – there can be no differentiation amongst team members in terms of performance ratings; they should all receive the same rating regardless of individual performance or contribution.

A comprehensive remuneration package

The Agile organisation calls for a more flexible view on rewards and benefits. Employees with scarce and critical skills look for more than competitive pay when joining or deciding to stay with an organisation. A comprehensive remuneration package includes salary as a 'ticket to the game'. It is the rest of the package that often seals the deal.

There have been some exciting developments in the area of remuneration packages and organisations have started introducing creative ways to reward individuals and teams. Elements of a comprehensive reward package are described below.

Competitive salary

As a general rule, the salary for your most in-demand positions should meet or exceed the market standard. This is especially important when you consider that bonuses are less frequently awarded in the current climate, so your initial base offering should be high enough to cover this. It is also important to review your salary and benefits structure regularly so you continue to offer competitive wages and attractive benefits for new and existing employees.

Flexible working hours

Demands on Agile teams are tough, so it is crucial they find a good work-life balance in order to remain happy at work. Many organisations are allowing knowledge workers to manage their own time, focussing on output and not desk-time. During Agile sprints, teams can work long hours and if they are not given the opportunity to have flexible working hours; if the pace and volume of the work can lead to stress and burnout.

Flexible working locations

Flexible work does not just relate to working hours. Employees should be given the opportunity to work remotely, from home, in coffee shops or wherever they feel most productive. The challenge for Agile teams is to remain sufficiently co-located to enable collaboration. This can be done using technology – video and audio conferencing allow virtual teams from all over the world to collaborate on a daily basis. Although co-location is of great benefit to Agile teams, you should not make working in a set environment a rigid condition of the job. You will often find that cohesive teams choose to sit and work together, but the results are much better when this is their choice, not a management decision.

Relocation packages

If your goal is to attract talent from overseas, you might like to consider adding a relocation package to your offering. Covering the cost of flights and arranging initial accommodation and basic relocation assistance can make the transition far easier for the candidate and boost your employer branding in the process.

Better healthcare benefits and coverage

One of the biggest rewards an organisation can offer right now is great healthcare coverage, including life insurance. Information is power in this area. Even the smallest of companies can reach out to their insurance or healthcare providers and ask them to quantify how much the organisation is paying and providing individually for an employee's healthcare benefits. Incorporating this important information as part of a total rewards statement gives each member of your team a fuller understanding of their total benefits and the value the company is contributing to their welfare. It is often an eye opener and one of the most valued job benefits you can provide.

Internal mobility and skills development opportunities

Agile methodologies are bringing the concept of craftsmanship back into fashion, as the ability to develop and master a craft in the new world of work is an indicator of growth and success. Traditional career paths are no longer sustainable in organisations as flatter structures and reduced levels of hierarchy make climbing the proverbial corporate ladder increasingly difficult. The way to progress your career is thus focused on either deepening your proficiency or skill level, or moving laterally and becoming multi-skilled. Remuneration packages that allow for skills development and mobility within the organisation are seen as more compelling and attractive for employees.

Personal days

These are different from vacation days or annual leave. Everyone needs some time off to take care of everything from medical appointments to household matters, or anything else that requires special time away from the office. Employees should not feel guilty about asking for and using these days. Give them the gift of a more balanced life and consider how you might be able to be more flexible in this area. Everyone will appreciate it even if they do not use the days, as they know they are there as a safety net.

Employee recognition programmes

According to Marantz Research, 58% of workers seldom, if ever, receive a "thank you" from their boss for a job well done, yet employees "who are recognized on the job are 5 times more likely to feel valued, 7 times more likely to stay with the company, and 11 times more likely to feel completely committed to the company".[155] Start regularly adding "Thank you" to your vocabulary. Be sure to also find new, visible ways to publicly acknowledge breakout work by your team. While you are at it, think about small ways of saying thank you, from gourmet coffees brought in to informal office catered lunches, or lunch at a local restaurant which is a team choice.

Unique benefits

Aside from meeting the industry standard, your remuneration package should aim to include some unique benefits that make your organisation stand out. Wellbeing and employee health initiatives, for example, are growing in popularity and have the benefit of conveying the message that you care about your employees' wellbeing, while also promoting productivity by encouraging workplace wellness.

FAIRNESS AND TRANSPARENCY

Fairness and transparency are critical components of team reward systems, especially in organisations adopting Agile methodologies. You do not have to publish individual salaries, but you should consider communicating with your staff how their pay rates are calculated. For example:

*   Do you pay more than, less than, or roughly the same as the going market rate?
*   Are certain roles paid higher than the market rate? If so, why?
*   Do you review pay rates? How often, and using what criteria?

There are varying grades of transparency that can be applied to rewards. Some examples include:

- Publish all individual salaries.
- Introduce a salary formula (considering role value, experience, loyalty, and location).
- Publish pay grades for reference roles (e.g., Senior ScrumMaster).
- Provide people with their individual salary band and position within that band (e.g., 85% to the midpoint).[156]

Reward systems that are not perceived to be fair and transparent will quickly demotivate your staff. Showing people that you have nothing to hide will help them realise that you are trying to be as fair as possible. Friederichs contended that "base compensation never motivates; it only satisfies or demotivates".[157] Although almost everybody would like to earn more money, what counts is whether they perceive their pay as fair. With social media and websites like Glassdoor, employees are able to gauge what the market will pay for skills and experience, and talent with scarce and critical skills know they are in high demand. You cannot afford for your staff to feel that their pay is unfair or that they are being exploited.

When looking at the fairness of pay, both distributive and procedural justice should be considered. Distributive justice considers whether an employee believes their pay is fair when compared to others, while procedural justice is when an employee believes their pay is fairly determined. Both of these are important to ensure that employees are engaged, productive and keen to stay with the organisation.

There are additional factors that employees consider when gauging fairness. Organisations that are large and have high brand equity are expected to pay more, whereas start-ups and smaller organisations get away with paying lower levels of compensation if employees feel they have a stake in the future success of the organisation. Another factor that is being increasingly considered is that of social relevance. Organisations that have an overt social purpose and demonstrably serve the greater good can also pay lower salaries without jeopardising employee perceptions of fairness.

No discussion on fairness would be complete without raising the issue of pay disparity linked to gender, ethnicity and disability. Equal pay for equal work is the notion, and in some countries the legislation, which prescribes that individuals who perform the same work should receive the same pay. Yet there are still widespread practices of women, specific ethnic groups and people with disabilities being paid less. Not only is this poor practice from a moral and ethical perspective, but it also carries legal and reputational risk.

UNLOCKING THE INTRINSIC MOTIVATION OF KNOWLEDGE WORKERS

The role of the manager in an Agile organisation needs to change to that of facilitator and enabler, as managers are no longer considered to have the highest levels of expertise or knowledge. This is especially true when 'managing' knowledge workers. With the abundance and accessibility of knowledge on the internet and the tremendous amount of learning and development available, there is no way for managers to keep one step ahead of their staff, nor should they want or need to. Knowledge workers in an Agile setting perform at their best when they are given autonomy, meaningful work and are allowed to manage themselves.

Research shows that extrinsic motivation such as pay drives performance in jobs that rely heavily on mechanical skills. Success in the knowledge economy requires a more cognitive skillset with creativity, conceptual thinking and the ability to deal with complexity. Employees with these skills rely on intrinsic motivation and are not sustained by a pay-driven reward structure.[158]

Pay can cause a lack of motivation if it is not competitive or perceived as fair, but after a point, money is no longer a motivator – intellectual freedom and self-actualisation become the driving forces of performance. Agile leaders need to understand that ideation, innovation and engagement are not motivated by money. In addition, you cannot get the best out of knowledge workers using fear, threats and intimidation.[159]

In this respect, leaders and managers need to have courage. Courageous leadership is not just about what you do as an individual leader; it is also about the amount of trust and autonomy you give to your employees. For many traditional leaders, it is far more difficult to trust their knowledge workers to self-manage and let go of command and control leadership styles. In Agile methodologies, the role of the leader is to drive alignment. The higher the level of alignment, the more autonomy the leader can give the employee.

Creating an environment of mutual influence is difficult, but ultimately leads to employees feeling heard and respected. Leaders can create this kind of environment by encouraging employees to:

- disagree where appropriate;
- advocate for the positions they believe in;
- make their needs clear and push to achieve them;
- enter into joint problem-solving with management and peers; and
- negotiate, compromise, agree, and commit.[160]

THE DEMISE OF THE ANNUAL BONUS

Agile remuneration structures are seeing the decline of using annual bonuses. In fact, annual bonuses are considered 'anti-Agile' and there is a call to abolish them completely. Any formal performance management and assessment system that provides feedback and reward up to 12 months after the event cannot be considered Agile in any way. Agile practices require feedback today for something that was done yesterday so that performance can be improved tomorrow.[161]

Six reasons why annual bonuses should be abolished

1. **Annual bonuses lag the behaviours that employers are rewarding**

   Agile organisations use short cycles to create business value. Waiting up to 12 months to evaluate, then reward, employee behaviour is anti-Agile. Imagine performing a work effort in the first three months of a twelve month cycle, and wondering for nine months how valuable that work effort really was to your organisation? By the time you receive your reward, you have lost the opportunity to maximise that behaviour over nine months (or, change your behaviour if your work performance was not as valuable as you thought it was).

2. **Annual bonuses are an imprecise reward**

   Think of all the stuff you do over 12 months. When you get an annual bonus, for which of those hundreds of work efforts are you being rewarded? Very likely, some behaviours you exhibited were stellar—and some were probably less-than-stellar. Annual bonuses blur the distinctions between highly valuable contributions and less valuable, or even harmful, contributions.

3. **Annual bonuses encourage competition and discourage collaboration**

   Organisations that award bonuses for individual performance have one, big pot of money to divvy up among all their employees. These organisations want to reward their outstanding performers for their great contributions by awarding them a larger slice of the bonus pie. Employees compete against one another to win one of the coveted "Outstanding" ratings from their employer so they get that larger bonus award. "Outstanding" employees are incentivised to continue standing out from the crowd; they have little reason to help lesser-performing colleagues perform better (altruistically, that might still happen, but not because most organisational bonus programmes encourage that behaviour). This is anti-Agile and a major organisational dysfunction.

   An Agile organisation needs everyone to continuously improve, and that requires higher levels of collaboration among all employees. Annual bonus programmes for individual performance do not incentivise collaboration, they incentivise competition and knowledge protectionism.

4. **Annual bonuses are insufficiently linked to individual performance**

For many employees, their individual performance is poorly correlated to their annual individual bonuses. Instead, factors outside their control strongly influence their annual awards. Although many bonus programmes ostensibly reward individual performance, such programmes are too often correlated to organisational (not individual) performance, whether someone was working on a highly-visible project, and who their performance-rating manager happens to be for that year — all factors that are outside an employee's control.

5. **Annual bonuses inhibit creativity, risk-taking, experimentation, and innovation**

Agile organisations seek creative, innovative solutions. Innovation requires experimentation and risk-taking, and, inevitably, failure. Being Agile allows for that. Agile cycle times are short (and some Agile teams deliver value via continuous flow). Failing fast is part of what Agile organisations do as they experiment and learn new ways of innovating together.

But, if creative, knowledge-based workers must protect their annual bonuses, they will innovate less, take fewer risks, and avoid the potential for failure. Bonuses that ostensibly reward individual performance and outcomes — not how those outcomes are achieved — do not reward the journey; rather, they only reward successfully crossing the finish line.

6 **Annual bonuses do not motivate an intrinsically motivated workforce**

Chris Alexander, co-founder of AGLX Consulting in Seattle, recently explained what knowledge workers really want in a LinkedIn discussion post:

"In knowledge industry work, which is intrinsically motivated, people want to engage in meaningful, challenging work. To do so, they need to feel that their compensation is fair and equitable. When the pay is equitable, additional pay does not result in increased performance (or *better* problem-solving). When knowledge workers don't feel adequately compensated (especially in today's competitive market), they simply leave and go work elsewhere."

Agile organisations pay outstanding employees their marketplace value without incenti-vising them to hoard their knowledge. They also create a collaborative working environ-ment where outstanding employees are not threatened if they make everyone around them outstanding, too.

(Adapted from Davis[162])

Compensation should be decoupled from an annual process to allow for a more flexible schedule. Where bonuses are still considered a necessary part of an overall remuneration package, there are a number of different approaches that can be used to distribute a bonus within a team. Some teams split the bonus based on seniority level and/or role, while others hand out equal amounts to each team member. Some teams pay out all the money directly, while others hold back a certain portion for a team event.[163] The way in which bonuses will be divided within a team should be discussed and agreed by the team before the work begins. In this way, all team members will be aligned if they receive a bonus, and team cohesion will not be threatened by differences of opinion about bonus distribution.

(CASE STUDY) **Patagonia**

Patagonia has eliminated annual raises for its knowledge workers. Instead the company adjusts wages for each job much more frequently, according to research on where market rates are going. Increases can also be allocated when employees take on more difficult projects or go above and beyond in other ways. The company retains a budget for the top 1% of individual contributors, and supervisors can make a case for any contribution that merits that designation, including contributions to teams.[164]

(CASE STUDY) **Rent the Runway**

The online clothing-rental company, Rent the Runway, dropped separate bonuses, rolling the money into base pay. CEO Jennifer Hyman reported that the bonus programme was getting in the way of honest peer feedback. Employees were not sharing constructive criticism, knowing it could have negative financial consequences for their colleagues. The new system prevents that problem by "untangling the two".[165]

CONCLUSION

Employees who thrive in Agile methodologies are driven by mastery, autonomy and purpose. Reward practices must align with Agile principles if they are to successfully drive the right behaviours. Using traditional reward practices that were designed for individuals can be divisive and destroy team cohesion and performance. Instead, team reward practices must be built to successfully enable Agile methodologies. Although this is an emerging practice, there is evidence to suggest that we are on the right track. Understanding new ways of motivating, recognising and rewarding teams, and specifically teams of knowledge workers, is one of the critical keys to success in adopting Agile methodologies. Not only does this form a large component of the new talent contract, but it also contributes to creating a highly desirable place of work.

CHAPTER 8

HUMAN CENTRED DESIGN – THE ART OF DESIGN THINKING IN REIMAGINING PEOPLE PRACTICES

Lizette Bester

INTRODUCTION

What is design thinking?

Seeing employees as customers and focusing on the employee experience.

A way to reinvent traditional people practices.

The necessity for empathy in design.

The use of MVP (minimal viable product) in challenging old-fashioned practices.

Over the years I have been amazed at how many Human Resources professionals forget about a critical part of their role. It is inherent in their job title yet seems to become blurry when faced with all the challenges our business partners leave in our virtual in-baskets. That element, of course, is 'human'; human beings are the reason why HR exists. Humans make up the clients with whom we interact on a day-to-day basis, yet we tend to factor them out of our equations when the going gets tough. We put in place policies, processes and plans that often aim at ticking the boxes, with the unintended outcome of this approach being that our crucial customers end up ignoring us or simply finding ways to circumvent us totally.

Luckily, more and more organisations understand that being 'Agile' is critical as the speed of change determines survival. Staying relevant in an ever-changing world is what will propel us into the future.

I have, however, been fascinated with the principles of Human Centred Design for a while. What astounded me was that it has roots in the IT and design spheres. IT is often more about the machines or the codes that make them run than about the end users, while design can easily focus on the product and not the consumer.

As HR professionals, we can learn a great deal from this approach to design.

WHY DOES IT MATTER?

Organisations need employees who are enthusiastic, engaged and excited. Whether designing policies, processes, systems or any other initiatives, we need to keep in mind that people want to feel like they have purpose and that they matter.

In his book, *The Power of Design*, Richard Farson stated that, "Design, the creation of form, has the power to transform culture, ignite education, foster community, and even broker peace. Design achieves its power because it can create situations, and a situation is more determining of what people will actually do than is personality, character, habit, genetics, unconscious motives or any other aspect of our individual makeup". He went on to say that, "Nobody smokes in church, no matter how addicted".[166]

Whatever we design we need to understand that it will impact either positively or negatively on the people we design for. We also need to remember that our reach goes beyond the existing staff to touch job applicants and service providers. Often, unintentionally, we design processes and interactions that leave people with a negative impression of the entire organisation.

Many organisations understand that the Employee Value Proposition (EVP) is a journey that reaches from recruitment to exit. We must ensure that each of the points that are crucial or 'moments of truth' leave employees with a sense that they matter and that they form an integral part of the ecosystem. If we fail in any of those moments, we run the risk of alienating and losing valuable resources as they will look for an employer who understands them and their value.

McKinsey defines a moment of truth as "an instance of contact or interaction between a customer and a firm (through a product, sales force, or visit) that gives the customer an opportunity to form (or change) an impression about the firm".[167]

For HR, the moment of truth acknowledges that our customers have many faces but that the staff who we employ makes up the bulk of our customer base. They are part of why we exist, and we need to ensure that we engage with them in a meaningful and valuable way.

By applying design thinking to our work we can fuel employee engagement and productivity while allowing creativity to drive innovation.

SO, WHAT EXACTLY IS HUMAN CENTRED DESIGN (HCD)?

Simply put, HCD is an approach to design that places people at the centre of any solution and ensures that what is put into operation is effective and meets the needs of the stakeholders it is intended to serve. It is an approach to design that assigns meaning to the people who will ultimately use the policy, process or system.

I love this quote from Harper Lee in *To Kill a Mockingbird*: "People generally see what they look for and hear what they listen for." As HR professionals, we need to learn the art of active listening to ensure we design processes, policies and other initiatives that serve both the business and its people.

HCD essentially has three phases:[168]

1. **Inspiration phase** – during this phase you learn directly from the people you are designing for. This means that you will need to embark on a journey of learning about these people to ensure that you fully understand their needs.

2. **Ideation phase** – once you understand then you make sense of what you have learned, identify opportunities for design, and prototype possible solutions.

3. **Implementation phase** – now is the time to bring your solution to life and introduce it to the stakeholders it is intended for.

WHERE CAN WE APPLY IT IN HR?

Josh Bersin at Deloitte predicts that HR of the future will stop designing "programs" and instead design integrated, high-value "experiences" that excite, engage and inspire employees.[169]

To achieve this, we can apply design thinking to the following:

1. **Organisational design:** this is a crucial HR function but one we often delegate to line managers. They tend to think bottom line and profits and our role in HR is to apply design thinking and its principles when we restructure or redesign roles and the organisation.

2. **Engagement**: design thinking can help us to devise ways to make work easier, more efficient, more fulfilling, and more rewarding.

3. **Learning**: this involves using technology and processes that ensure learning experiences are designed for people and that their needs are met either through the 'how' or the 'when' they access learning and how the learning is structured.

4. **Analytics**: for many years we have talked about HR Analytics and Metrics, but design thinking ensures that we do not just use the data to look good on the annual report, but that it is linked to ensure that the EVP and employee experience is enhanced.

5. **HR skills**: design thinking must be learnt to ensure that a wide range of solutions are used to incorporate an understanding of digital design, mobile application design, behavioural economics, machine learning, and user experience design.

6. **Digital HR**: in the new world of work this is crucial as we need to develop new digital tools that can make work easier and better.

BUT WE DON'T HAVE THE LUXURY OF REDESIGNING EVERY HR POLICY, PRACTICE OR INITIATIVE!

The lack of time and resources and the ability to 'start from scratch' is not unique; all organisations I have ever interacted (or worked) with have the same problem. But this does not mean that nothing can be done. Starting with intention and then making incremental changes over time can be very effective. By applying design thinking to everything you do from a certain point in time can start to embed the approach and help employees understand that the organisation has 'heart'.

Many of the most effective and respected organisations world-wide take a very determined approach. Examples of such intent can be seen through:

*   Cisco[170], which hosted a non-tech hackathon to explore a wide range of HR issues with its employees. From this they identified 105 new solutions for its global workforce of 71,000. This ranged from experiences in recruiting to on-boarding, and learning and development. To delight employees, Cisco has identified "moments that matter" – such as joining the organisation, changing jobs, and managing family emergencies – and redesigned its employee services around these moments.[171]

- AirBnB[172] has changed the Chief HR Officer function into a Chief Employee Experience Officer function, recognising that "experience" is the essence of a workplace, especially among millennials.[173]
- Pixar[174] has an Employee Experience Manager who provides outreach, consultation and support to a variety of groups and individuals. This means lots of face time and conversations with employees and managers to better understand their experiences, challenges, and development needs.[175]
- The golden thread across all these initiatives is interaction. Touching base and knowing your people is key to design thinking, as is empathy.

SO WHAT IS EMPATHY REALLY?

Emi Kolawole from the Stanford Design School says that design thinking is all about empathy.[176] This links back to my Harper Lee reference when she says: *"You never really understand a person until you consider things from his point of view... Until you climb inside of his skin and walk around in it."*[177]

Let me explain this by way of a personal example. I was once asked to design a 'Loss of License' policy within the aviation space. As I was reasonably new to the environment I had no sense of how crucial and controversial this would be. Ignorantly I started with the existing (old) policy and then reviewed the new insurance documents and checked with the medical team on the ways in which the processes within the policy should interact. Arrogantly I started writing a lovely (well I thought so) policy document in record time (I may add). With that same arrogant bravado, I took the policy for review to the Executive within the Flight Operations area expecting to get a pat on the back and a thumbs-up. What I did get was neither of those and in fact walked out of his office feeling shell-shocked. Because I did not take the time to engage with or explore with all the stakeholders, I had absolutely no sense of the highly emotive and sensitive nature of this policy.

The reason why this was such a critical policy was that it related to a pilot's ability to earn a living. Although a big part of this policy related to insurance principles and the processes required to evaluate the eligibility to receive benefits, the actual drivers of this policy were far more human. In almost all the cases where a pilot lost his/her licence it was related to a change in their health and general wellness, which comes at a high emotional cost. Add to this the fact that they can no longer fulfil the role that they train and fly countless years for, and round it off with a formula for payment they will receive as a result, and you have a highly explosive mix.

Only by applying design thinking to the process and by engaging not just with insurance departments and medical professionals but also with those who were the beneficiaries (recipients) of these benefits, could the policy and processes be refined. By speaking to the pilots and understanding their emotional perspectives, as well as the value they derived not just financially but also personally, I could design a policy that would address their concerns as far as possible. By doing that I understood that this policy was more than a simple set of rules; it had embedded in it the intangible message of value. It was a way of saying that we regretted that they had lost the thing that defined who they were professionally, and that although money cannot replace it, it could in some way compensate them for this great loss.

OLD VERSUS NEW

We need to accept the need for problem solving not just with the existing, but also the new, challenges faced by the business. By having an open-minded approach and challenging existing limitations and influences, we can start changing behaviours and the culture within the organisation.

Design thinking can be adopted and applied to existing practices through incremental innovation or defined afresh through the introduction of novel ideas and solutions which arise from interactions with people. Sometimes disruptive innovation is required to ensure real changes are achieved.

There will be those who are opposed to innovation and take a view that, "This is how we have always done it here", but by wide engagement and a consistent application of design thinking, as HR professionals we can ensure a change in approach and ultimately organisational culture.

By asking questions around issues including the physical, perceptual, cognitive and interactive requirements of employees, we will give ultimate meaning to our policies, systems and services, which will create a comfortable and productive space for everyone in our business.

Some of the questions we need to ask when reviewing or designing anew are:

1. Who?
2. What?
3. When?
4. How?
5. Why?

If we can get satisfactory answers to these questions we can start to evaluate whether a policy, process or system is really required, and if so, why we need it. This will also help us design and refine to give everything we do in HR a real purpose and value.

MINIMUM VIABLE PRODUCT (MVP)

Before exploring the design principles in HCD in more detail, it is important to look at the practice of MVP.[178]

MVP was defined by Eric Ries, a consultant and writer on start-ups, as a development technique in which a new product or website is developed with sufficient features to satisfy early adopters. The final, complete set of features is only designed and developed after considering feedback from the product's initial users.[179]

Applying this to HR makes sense. Often we are frustrated by the fact that we design and develop HR initiatives, policies or practices, just for them to stay on the shelf for years and for them to be discarded later, or even worse, to be introduced but never used.

By combining design thinking and MVP we can design and develop for the future, however MVP requires three elements:

1. Enough value that people are willing to use it initially.
2. Demonstrates enough future benefit to retain early adopters.
3. Provides a feedback loop to guide future development.

Critically, early adopters can see the vision or promise of the final product and provide the valuable feedback needed to guide us forward. This is especially useful for technically-orientated products like e-learning, self-service or other technology-based HR solutions. Within the Agile organisation, MVP as a project approach is defined as the simplest thing that could possibly work. I recently heard someone say it is applying the low-tech solutions to the high-tech problems. Simplifying and ensuring that solutions are easily understood and applied is key within the Agile context.

By placing the simplest version of a solution into the hands of the users, you make sure that you are on the right track. You can also categorise the highest priorities for the next phase of your project.

MVP is the "first draft" and it must meet the threshold of a successful iteration.

A real first draft must be able to:

a.   do real work;
b.   be able to be evaluated; and
c.   actually be evaluated.

For the first draft to be evaluated, someone must use it — and provide feedback. That feedback must be gathered, studied, and, most importantly, used to improve the next iteration.

The MVP has to be a reasonable stand-in for the actual/final solution. This is crucial in our view as it is easier to make a change to a first draft than to a final version. In HR we often believe we should implement and be done with it; MVP helps us to draft and then refine before we can say it is the final version.

Using MVP is also a very cost-effective way of managing any HR project. We can assess value prior to spending unnecessary money or other resources. If it is not effective or required, we can simply opt to discard prior to spending excessively and having fruitless and wasteful spending.

WHERE MVP AND HCD CONVERGE

To define the project scope, whether big or small, the design team needs to assess and set out the project scope and deliverables very clearly. But getting there needs to be preceded by the HCD principles. Let's just get back to these for a bit, as we cannot define the deliverables without each of the phases:

I.   Inspiration phase

The inspiration phase requires the developer (for our purposes, HR) to actively engage with the users of the policy, process or initiative to understand them fully. This is crucial to ensure that we get a sense of what drives these stakeholders. Only by understanding what employees value, what their emotional drivers are, and what motivates them, can we even begin to design with purpose.

The inspiration phase requires us to engage with not just some, but with all, the stakeholders by speaking to them (even if it is electronically). This will help us understand their emotional perspectives, the value they derived personally, and how we should engage with them around this. By doing this we can have empathy and build solutions, processes and initiatives that will have real impact.

II. Ideation phase

This phase is about actually forming ideas and coming up with solutions based on the feedback in the inspiration phase. There are no hard and fast rules around ideation, but generally we can benefit from:

i.   clearly identifying the problem/obstacle or objectives; then
ii.   brainstorming or crowdsourcing; and then
iii. streamlining/pairing-down or refining.

Once you have refined all the ideas and potential solutions, you can set the project plan and work towards developing the solution that will be your MVP.

III. Implementation phase

This is the real convergence of HCD and MVP. Implementation does not mean you are putting in place the final version; this is simply the description of a process of which MVP is a crucial element. Testing your solution and then getting feedback not only ensures that employees are on board, but also that your solution is practical and viable.

Megan Torrance, chief energy officer at TorranceLearning, put it beautifully when she said: "Agile project management... encourages working out loud, publicly, collaboratively."[180] This is a view that resonates with many HR practitioners as we often have to deal in 'Chinese Whispers' and guestimations because we are too worried that if something does not work we will lose credibility. In this way we openly state we do not know if something will succeed, but that we are open to finding a solution that will.

During the implementation phase (and MVP), user testing is critical. Do not hold back on this phase. Get your test teams in place and guide them on the types and forms of feedback that will assist in refining and finalising the project. This seems time consuming, but you have to ask yourself – if you don't take this time, won't you have wasted all the time invested in the project?

Practically applying this approach within your business will require a full understanding not just of the process of user testing, but also who will make the best testers of the prototype. By having the right people involved in testing the feasibility and accessibility of the policy, process or HR product/system you have designed, you will get feedback that will assist in effectively modifying and aligning according to the user's needs.

CHALLENGING OLD THINKING

For many of us within HR this approach is quite new (even if it makes perfect sense). It assumes that as HR we are trusted partners who have credibility and the ability to really influence business decisions. This may not always be the case, however. Sometimes we are the makers of our own fate. I have heard on more than one occasion that HR has no concept of the business or the challenges faced by the staff within the business.

We may have to start with ourselves. Some of the ways we can begin to change perceptions include:

i. **Get to know the business and your people**: understanding your business at all levels is critical. It is a double whammy as you will then also get to know the people who work within areas, their specific challenges and the operational imperatives that drive them. If you are in retail, go work in a shop for a day or two. If you are in hospitality, go face the customers. If you are in manufacturing, go and work on the various shifts and look at the challenges within the factory or plant environment. By doing this you also become visible and show a willingness to understand the business environment.

    I spoke to someone in operations recently at a regional office of a large construction concern. In almost six years with the company he has not seen a single person from HR. His view of HR is that they are simply a nuisance who are constantly impeding his ability to deliver on his objectives. They do not understand his pressures, the types of individuals who will fit into his environment, or the challenges they face as a satellite office. This is a sad state of affairs and means that HR becomes disposable.

ii. **Know the company strategy and business objectives**: HR cannot do what it likes. True HR departments are not just payroll and HR admin (which obviously is crucial), they are enablers and a key part of delivering on the business strategy. All HR initiatives and projects should be aimed at helping the business, through its people, to achieve and deliver on its promises. By knowing the business strategy, the HR strategy can be aligned and everything we do can be targeted. In this way we become a trusted partner and no longer just a cost centre.

iii. **Be proactive**: by knowing the business objectives and having an aligned strategy, HR can plan and put solutions and processes in place that are ahead of the curve and ready for the business when it needs them. Practically this means that if HR is in touch with the business as a trusted partner, they will know about projects or challenges on the horizon. In this way, plans, systems, policies or processes can be put in place, or at least designed, to ensure that there are minimal disruptions and that the design process is not rushed for the sake of delivering 'something'. By being proactive, HR can deliver practical, fit for purpose solutions. If we are simply reactive we constantly play catch-up, and it feels like we are running behind picking up the pieces. As HCD is not a quick fix solution, by being proactive we can pre-empt business needs and start work before it becomes critical.

iv. **Develop our own skills**: Effectively managing projects with HCD and MVP at their core and really creating experiences that will excite and inspire staff requires us as HR practitioners to develop not just our HR knowledge, but also our other proficiencies like project management, facilitation and consultative skills. Without these we will struggle to pull together resources while making sure we deliver on our promises.

v. **Be brave!**: In some cases, we will need to be brave. This bravery is not bravado. We may need to take tough decisions and make choices that are not popular.

Changing perceptions and getting buy-in for a new and innovative approach obviously takes time and a considerable amount of effort, so we need to be tenacious in our approach and make people understand the value of designing with people in mind. Not every bit of work we do in HR will require a song and dance. Some things will be obvious and a new approach may not be disruptive, however, especially in a business with a secretive culture (and sadly there are many of those), we will need to convince not just the leadership but also the staff of the bona fides of this approach.

As HR, we will have to ensure the following:

- We sell the approach to senior leaders. Without their support we will be barking against thunder and our projects will not have the priority they require. This may mean that we need to have our business hats on when we engage with these leaders. Showing real value on investments or potential upside is key, as they are operators who have a business objective and they will only support us if we can show how we support them.
- We create safe spaces for all the stakeholders. Particularly in the inspirations phase, it is crucial that we create safe spaces for those who are willing to participate. A clear understanding of them as individuals or as a group requires us to park all our perceptions and our thinking and listen to them with intent. Protecting individual views (which may be controversial) is also important so that we have a full view of the stakeholders in order to create ideas and solutions that will fit with them.
- Empathy, empathy, empathy... empathy! Not just for the people who will be using or applying our solutions and initiatives, but also the leadership who need to

manage the people. As HR we are best placed to demonstrate empathy while still being realistic and supportive.

- Practice what we preach. If we can show that we are consistent in our approach, whether reviewing the existing or creating the new, we will gain credibility. This takes time, but repetitive consistency will turn the ship. Some ships turn slowly while others are Agile and quick to assimilate new approaches. Knowing your culture, leadership and staff will assist you to have reasonable expectations while still modelling what good design can achieve. Knowing the practicalities within your environment will ensure that HR delivers solutions that meet the business requirements.

HOW DO WE APPLY THIS PRACTICE?

Practically, this means that when we embark on any changes or introduction of new initiatives within HR, we need to clearly assess how and why we are doing it. We will then need to ensure that we clearly map the approach and processes, as well as how we define success.

If we are about to change an existing policy, process or approach, we need to ensure that we engage in an appropriate way with the various stakeholders and that we make changes and amendments to meet their needs. Sometimes this may require being brave. If a policy needs to be discarded, then be brave enough to do so. If we need to admit that we were wrong, then let's do that.

When we embark on designing, developing and introducing a new solution, we need to ensure that it is not viewed as a fad but that it is driven by a real business rationale. Let's ensure that we do not just do things because it is the perceived 'flavour of the moment', but because it is part of a well thought-out strategy that is business relevant.

Honest and open engagement with our customers will set the tone, help us with credibility, and ensure that when people engage with us that we are considered trusted and reliable partners. By taking a systematic approach we also show that we are business-minded and that we can produce solutions and experiences that are consistent with the business strategy, while supporting not just the warm-and-fuzzy part of our HR mandate, but also our business partner role.

In my view, HCD is not a nice-to-have within HR, but an imperative that will allow us to work with business to achieve results while ensuring that the EVP is a consistent and pleasant journey for all who pass through our employment doors.

CONCLUSION

Human Centred Design (HCD) is an approach to design that requires us to engage with our stakeholders in three phases:

1. Inspiration Phase
2. Ideation Phase
3. Implementation Phase

Embedded in HCD is empathy, which forms the foothold of the approach and allows us to keenly understand those we are designing for.

During the implementation phase it is valuable to use the Minimum Viable Product (MVP) approach to Agile Project Management. This allows us to place the simplest version of a solution in the hands of the users to ensure that they are on the right track. It also allows them to categorise the highest priorities for the next phase of your project. As HR, we will need to:

- get to know the business and our people;
- know the company strategy and business objectives;
- be proactive;
- develop our own skills; and
- be brave!

We will also have to:

- sell the approach to senior leaders;
- create safe spaces for all stakeholders;
- show plenty of empathy; and
- practice what we preach.

Good luck in applying these principles and approaches! The upside is that you will work within an environment where your work is relevant, the solutions practical, and the outcomes amazing. People will not only feel valued, but they will engage with solutions and experiences in a whole new way. HR will become a trusted partner with credibility and real business appreciation.

CHAPTER 9

WORKSPACE DESIGN AND CO-LOCATION

INTRODUCTION

The physical space in which we work makes a significant difference to how we *feel* when we are at work. The lay-out and design of the space, together with the perks/ benefits offered, can positively contribute to the employee value proposition and also impact on innovation, productivity and engagement. Agile methodologies rely heavily on collaboration, and although co-location is not essential, it certainly fuels and enables collaboration. This chapter focuses on the benefits of using workspace design and co-location to enhance the adoption of an Agile mind-set and Agile practices, as well as generally improve the employee experience.

THE IMPACT OF WORKSPACE DESIGN ON EMPLOYEES

The evolution of workspace design for most knowledge workers has resulted in the prominence of open plan offices. We have moved from individual offices, to large floors of cubicles, to large floors of shared desk spaces with reduced/no cubicle dividers. The *Gensler 2016 Workplace Survey* estimated that over eight million UK employees work in open plan environments, but many of these are not designed to promote creativity or innovation; 70% of employees are forced to work in the same place throughout the day, leading to only 33% of respondents feeling energised at the end of the day.[181]

While intending to promote collaboration, innovation and productivity, we are finding that traditionally designed open plan offices often result in just the opposite. Employees are feeling the negative impact of lack of privacy, frequent interruptions and disturbances, lack of ability to concentrate, high levels of background noise, and irritation with co-workers who have annoying habits like chewing gum loudly, sniffing or eating smelly food. Traditional open plan offices do not offer variety or choice, nor are they tailored to specific tasks and practices.[182]

Figure 9.1: An example of a traditional open plan office with well-defined cubicles[183]

Figure 9.2: An example of a modern open plan office where the use of cubicles/dividers is very limited[184]

A well-planned workspace can lead to collaboration, make it easier for people to complete tasks quickly and effectively, promote innovation, and have a positive effect on health, wellbeing, and engagement. In the new world of work, innovation and employee engagement can make a difference to whether organisations remain relevant and have the ability to attract and retain talent. In the *Gensler U.S. Workplace Survey*[185], a quartile analysis comparing the top 25% of innovation scores to the bottom 25% of innovation scores was conducted, with significant differences between the most innovative and least innovative groups being found. These differences included how employees work, what they do, the amount of time they spend at their desks, and even a difference in the amount of time they spend away from the office. Innovators were found to have more effective workspaces, better designed spaces, and twice as much access to certain amenities, leading to more choice, meaning and purpose.[186]

THE LINK BETWEEN WORKSPACE, WELLBEING AND ENGAGEMENT

Employee engagement is a well-recognised driver of discretionary effort, productivity and employee retention. Despite the enormous amount of time, energy and money spent on trying to enhance employee engagement, many employees remain disengaged. Aon Hewitt runs an annual survey to measure employee engagement in more than 1,000 organisations around the globe. Their 2017 survey results indicated that only 24% of all employees fall into the Highly Engaged category, while another 39% can be categorised as Moderately Engaged, putting the global engagement score at 63% compared to 65% the previous year. The results for Africa were slightly worse, with an engagement score of 61%.[187]

One important way in which organisations can help enhance employee engagement is to focus on wellbeing. Studies conducted by Steelcase (a world leader in workspace research and design) have shown that the physical work environment can have a strong impact on employee wellbeing and engagement. In 2014, Steelcase commissioned a study of 10,500 workers in 14 countries throughout the world. Their results clearly illustrate that employees who are highly satisfied with the places they work are also the most engaged.[188]

The table below depicts the results for highly-disengaged employees who were not satisfied with their work environment.

Table 9.1: Results from highly disengaged employees describing the impact of their workspace[189]

My work environment does not allow me to:	
85%	Concentrate easily
84%	Easily and freely express and share my ideas
85%	Feel relaxed, calm
57%	Physically move during the day and change postures
79%	Accommodate mobile workers
84%	Feel a sense of belonging to my company and its culture
87%	Work in teams without being interrupted or disrupted
86%	Choose where to work within the office, based on the task I am doing
59%	Move around easily throughout the day
65%	Socialise and have informal, relaxed conversations with colleagues

The link between wellbeing and workspace is multi-dimensional. Physical, emotional and cognitive wellbeing are all impacted by workspace and can all be enhanced using workplace interventions.

HOW TO DESIGN A WORKPLACE

Physical Wellbeing

In the past, the focus on work-space design and physical wellbeing revolved around er-gonomics. For example, it was considered essential to provide high quality office chairs with good lower back support because employees were expected to sit at their desks for long periods of time, working on desktop computers. Today we appreciate the importance of movement throughout the day, as changing one's posture stimulates one's mind. Steelcase's research shows that 96% of highly engaged workers are able to move freely and change postures throughout their day. From a design per-spective, the modern workplace should offer a variety of indoor and outdoor spaces that offer posture choices and encourage walking to create energy.[190]

Figure 9.3: Desks with adjustable heights are one of the workspace features that can be used to allow for a change in posture and improve physical wellbeing[191]

Emotional Wellbeing

One's quantity and quality of social interactions have a significant impact on one's emotional wellbeing. When employees are isolated or do not have enough quality interactions, they can become disengaged. This also makes it more difficult to collaborate, inno-vate, solve problems and be open to change. Technology and social net-working platforms have made it easier to interact, which is great for teams that are not co-located and are spread geographically. An unintended consequence, how-ever, is that we are increasingly using these platforms to interact with people close by. It is not uncommon to text or email a colleague who sits a few metres away from you. The workplace needs to encourage and enable employees to see and hear each other in formal and informal settings. This is essential not only for emotional wellbeing, but also for collab-oration and the building of social capital.[192]

Figure 9.4: A mix of formal and informal meeting and interaction spaces promotes quality interactions and emotional wellbeing[193], [194]

Cognitive Wellbeing

Research shows that 98% of highly engaged employees say they are able to concentrate easily at work, and 95% are able to work in teams without being disrupted. This is testimony to good workplace design. In most workplaces, the propensity to be disturbed and lose concentration is very high. In fact, our thinking is interrupted, on average, every three minutes. Even brief interruptions of just a few seconds cause us to make twice as many mistakes. After focused work is interrupted, it can take up to 23 minutes to get back into flow.

So much of our work requires thinking, problem-solving and focus that it is critical for employers to use workplace design to help employees manage the cognitive overload of their daily lives. A well-designed workplace can help reduce stress and promote cognitive wellbeing.[195]

Figure 9.5: Quiet spaces in an office environment allow for privacy and give employees a place where they can focus without being disturbed, promoting cognitive wellbeing[196]

ENSURING MAXIMUM COLLABORATION, INNOVATION, PRODUCTIVITY AND ENGAGEMENT

There are a number of evidence-based workplace design elements and best practices that can be used. Together, these will maximise your ability to foster collaboration, innovation, productivity and engagement at work.

Focus on multiple ways of working

There is no one optimal office design; instead the workplace should include a variety of spaces, modalities and environments in which employees can work. Also remember that working is not limited to sitting at a desk. The workspace needs to create spaces so people can focus as individuals and teams, as well as collaborate, learn, and socialise. Many organisations like SAP, LinkedIn, Cisco and Airbnb have an open environment with cubicles, collaborative innovation hubs, co-working cafés, conference rooms, smaller meetings rooms, and areas for presentations. The key is to shift away from having a single floor plan to integrating and incorporating multiple floor plans.[197]

The workplace is thus designed as an ecosystem which supports the physical, cognitive and emotional needs of people, and gives them choice and control over where and how they work. Not only is this a benefit in terms of the working environment, but the choice and control also serves as a powerful component of the employee value proposition, promoting autonomy.[198]

Make the space reflect the culture

The workspace of an organisation must match its culture. If there is a misalignment, the investment made in the workspace will have little return. One of the fundamental building blocks here is flexible work practices. While most knowledge workers are happy and used to working long hours without earning overtime, there is a growing expectation of flexibility. This flexibility applies to working hours, working remotely or from home, and managing one's own time and work-life integration. Building a modern workspace with state-of-the-art facilities and cutting-edge design will not achieve the desired results when employees are micro-managed, have to maintain rigid office hours, and are not provided with the autonomy to manage their own work life. Organisational culture is therefore an essential component in extracting value from workplace design.

Too many organisations make the mistake of installing artefacts in the absence of culture transformation and think they are now Agile. Buying a ping-pong table and bean bags and allowing staff to wear jeans to the office is not going to change the culture or encourage an Agile mind-set.[199]

Look at how employees work

Before investing in a workspace design or re-design, best practice suggests that employee habits and workspace utilisation be studied. In this way, you can make informed decisions about how to create a workspace that addresses the issues of concern. At Atlasssian, a study was conducted where sensors were fitted to employees' desks to find out how often they were used. They realised that their employees did not really spend that much time at their desks as they were often on the move – going to conference rooms, finding quiet areas, and working from other spaces.[200] Studies like these have led to the rise of 'hot-desking', where employees do not have a dedicated desk. Instead they have a private locker to keep personal or confidential material, and when they are in the office, they can sit at any available desk.

Hot desking has its pros and cons. On the positive side, office space and furniture is used more efficiently, and you can get to know colleagues with whom you do not usually work. Data privacy is also better managed as you cannot leave documentation or personal belongings on a desk. On the negative side, you may want to sit near the people in your team, but if you arrive at different times, adjacent desks may not be available. In some offices where flexible work practices are employed, they may install fewer desks than there are employees, counting on the fact that at any given time a percentage of employees will be working remotely. This works most of the time, but coming to work and finding all desks occupied is not an experience that engenders a sense of belonging.

Another advantage of conducting employee preference studies before starting a workspace design is that you will be able to determine what is considered valuable to your workforce. This can be influenced by generational differences, gender, psychographics and other factors. For example, employees may prefer a snack bar to a pool table, or couches to bar stools. If your workforce is large enough, chances are you will need to have a combination of all to suit diverse preferences, but you will never know unless you ask.

Incorporate environmental design

Green building practices which focus on environmental sustainability are highly sought after. Not only do these practices result in long-term cost efficiency, but they also support the value of sustainability. From a workforce perspective, many of these practices have benefits for employees and can promote wellbeing and engagement. There is a long list of environmental design elements which are considered 'a ticket to the game'. In other words, none of these is a differentiator, but all must be in place as basic and essential features. For example:

- Ventilation and heating/air-conditioning systems.
- Adequate lighting.
- Acceptable noise levels.
- Access to sufficient washroom and toilet facilities.

An element of workplace design that has been scientifically proven as beneficial in the environmental design space is the use of plants. The *Human Spaces Report* (2015) studied 7,600 office workers in 16 countries, and found that nearly two-thirds (58%) of workers had no live plants in their workspaces. Those whose environments incorporated natural elements reported a 15% higher wellbeing score and a 6% higher productivity score.[201] Other studies have shown similar results. Research conducted by the University of Exeter (2014) showed that employees' productivity increased 15% when work environments were filled with just a few houseplants. They asserted that adding just one plant per square metre improved memory retention and helped employees score higher on other basic tests.[202]

Figure 9.6: The use of plants in the workplace has been scientifically proven to improve productivity and wellbeing

CO-LOCATION AND THE ADOPTION OF AGILE METHODOLOGIES

Agile methodologies include small teams/squads as the base unit of production. These teams can be co-located or decentralised. Where possible, it is highly advisable to co-locate teams, as working in close proximity increases the immediacy of information and shared knowledge, thereby supporting speed in problem solving. Sharing the same physical space also allows for nuanced communication and nonverbal cues that enhance empathy and trust. Familiarity and rapport develops far more quickly in co-located teams and this increases the pace and quality of collaboration.[203]

When teams are decentralised, it can be challenging to ensure high productivity and efficiency. Delays can arise due to differences in time zones or non-availability of reliable communication tools.[204] While it may be costly to re-organise a workspace or move employees to different buildings/locations to co-locate, the benefits far outweigh the costs over time. It is often not possible to co-locate teams as a result of skills shortages, however. The reliance on contingent workers from a range of countries like India, Ukraine and Poland is common practice. In these instances it is necessary to facilitate communication, understand cultural differences, synchronise work, and foster knowledge sharing.

When embarking on an Agile transformation journey, the physical workspace can be an overt and powerful cultural lever. Reorganising a workspace gives employees explicit signs that things are going to be different; the workplace should look different, feel different and be a stimulus for new ways of working together. Many organisations have successfully used a workspace reorganisation to not only enable Agile practices, but to shift the culture and improve employee engagement. This is best demonstrated by the Steelcase case study below.

CASE STUDY

Creating a workspace for an Agile IT environment[205]

Steelcase is one of the world leaders in workspace research and design. They recently embarked on their own Agile transformation journey in their IT department. Their goal was to use their workspace to support Agile processes and foster a more Agile culture among the 400 IT professionals at Steelcase in Grand Rapids, United States.

Some of the differentiators that made the Steelcase case study stand out were that:

- the leaders and teams had an unprecedented depth of involvement;
- learning was conscious, intentional and codified;
- the amount of prototyping was significant; and
- the engagement processes were comprehensive.

They began with a workplace advisor study to discover how the IT teams were working and how they preferred to work to achieve their goals. They used sensors installed in the former workplace to measure and analyse how people were using their space.

Space study findings:

- IT spaces were underutilised. The individual workstations, small group and large meeting rooms were all occupied less than 30% of the time.
- Most collaboration was happening in small or mid-sized groups. People reported collaborating half of the time.
- Video display and acoustical privacy were in high demand.
- While only 30% of IT team members were designated as mobile workers, an additional 45-50% of people demonstrated moderate levels of mobility.
- About 20-25% of IT exhibited low mobility, including people who were away from their desks for less than 30 minutes at a time.

The next step was to conduct a culture survey to better understand the organisational culture in IT.

Culture survey findings:

- People said small teams worked very well, but larger teams were not as effective.
- Conflicting priorities and bottlenecks slowed progress and kept teams from working well together.
- There was an opportunity to increase trust among team members.
- People desired greater connections within the larger IT community.
- It was challenging to adjust to the ebb and flow of work that needed to get done.
- More mentorship was needed regardless of age. People wanted more opportunities to grow their skills and learn from others inside and outside of their discipline.
- A siloed work process kept IT from operating as a unified team.

These findings were combined with "day in the life" interviews, where researchers asked different IT team members to walk them through their typical work day. The collection and synthesis of all of this information helped the project team delve into people's challenges and expose what was working well and what was causing headaches.

Employee workshops were then conducted to help people create a framework for their future work experience. For example, IT employees wanted to encourage learning across the department instead of keeping knowledge in silos.

Agile user prototypes were built for iterative testing and refinement. These prototypes were built based on employee experience and were designed to inform the future IT space and support the culture change taking place. Groups representing each of the three types of IT teams moved into a basic space and began to experiment. Leaders empowered them to hack their space — grab couches, move desks, commandeer monitors — finding whatever they needed to do their best work.

Open forums like **town halls** and an **intranet blog** offered full transparency into what was happening, what was driving change, and what team members were learning.

In order to **enable Agile teams**, the following requirements were identified:

- Providing space within Agile studios for Agile ceremonies: Sprint, Sprint Planning, Daily Scrum, Sprint Review, Sprint Retrospective.
- Providing space within Agile studios for Agile artefacts: Sprint Backlog, Product Backlog, Increment Board, Information Radiators, Customer Personas.
- Designing adaptive space: giving permissions to teams to 'own' their space, rearrange furniture and self-organise around their processes/modes of work.
- Providing spaces for individual flow while staying in/near the studio, balancing the connection with the team.
- Promoting embedded learning: enabling team members to problem solve and build knowledge in real time.

All of the survey and prototype learnings were taken into account and a workspace was designed to specifically enable Agile practices and embody cutting-edge design excellence.

The designers developed four distinct areas for the new IT space:

1. **Neighbourhood**
    Agile teams reside in the overall Neighbourhood, which includes their studio and other flexible spaces. By living together, teams accelerate the flow of information and problem solving. Within the neighbourhood, you will find:

    Agile Studio – Owned and unique to each team, these function with a "kit of parts" that use common furniture and applications, but remain easily customisable and provide for high mobility with the space. Teams are empowered to move their furniture based on their needs. (Discipline teams also have their own spaces, called Discipline Studios. These are configured differently to fit their needs.)

    Front Porch – Touchdown spaces for resident teams; used for quick one-on-ones, stand-up meetings, customer touchpoints and more.

Flex Camp – Shareable space for IT staff who do not have a designated team space and business partners.

Community of Practice Centre – For special groups that need a short-term space to share knowledge or work together.

2. **Business District**

    The Business Districts are designed to facilitate collaboration within IT and also with IT's internal customers in the business. The space includes formal and informal places to meet. Some are reservable and some are touchdown.

3. **The Town Square**

    A series of larger, shared meeting spaces is a central anchor in the design. Between rooms are social spaces to encourage critical connections that support the development of trust within and between teams. The Town Square also includes:

    *   **Leadership Area** – This is a "nerve centre" for leaders to work in, which is designed around keeping leaders in close proximity to each other and their teams, as well as encouraging collaboration even at the highest level of the organisation. Both analogue and digital displays are   persistent reminders of what teams are working on in IT.
    *   **Café** – A place designed to encourage informal interactions among and between teams.

4. **Garden**

    The Garden area is a shared place for respite, heads down work, or short collaborations. The space is designed to bring biophilia into the workplace with plenty of natural light, views and greenery. One side of the Garden area was designed to support more collaboration, while the other side is focused on individual spaces.

The Front Porch is used for quick one-on-ones, stand-up meetings, customer touch points and more

The Agile Studio – owned and unique to each team

The use of Agile artefacts is common to all Agile Studios

Reservable meeting rooms in the Business District

CONCLUSION

Employees who work in an environment that promotes flexibility and autonomy and a workspace that offers a diversity of work modalities will almost certainly be more productive and engaged. There is no doubt that many of the world's most successful organisations are paying attention to the value of workspace design as they embark on their Agile transformation journeys. The co-location of teams enables collaboration, and when coupled with a workspace that facilitates Agile practices, you have a recipe for success. When executed well, workspace design can be an expensive exercise, but in the new world of work where we compete for talent and scarce and critical skills, organisations are going to have to consider this an investment in not only their people, but also their employee value proposition and culture.

ROBOTIC PROCESS AUTOMATION AND ARTIFICIAL INTELLIGENCE – THE END OF THE WORLD AS WE KNOW IT?

Bryden Morton and Chris Blair

INTRODUCTION

The idea of artificial intelligence has been around since the early 1900s, originally in the form of fictional writing and later seen in films. The minds of these writers and film producers imagined a world where the role of robots in society was elevated from the role of machines in their present society. These individuals imagined technological advances that would provide a machine with the ability to process sets of information and make decisions based on the information that the machine was taught to process. At this stage, the idea of creating robots and artificial intelligence was merely a dream, and was limited only by the imagination of those who pictured these characters and concepts. The task of taking these ideas and translating them into reality was significantly more challenging, as the constraints of reality proved to be a hindrance to the imagination of those exploring these concepts at the time.

Turing[206] published a paper called *Computing Machinery and Intelligence*, in which his opening proposal was: "I propose to consider the question, 'Can machines think?'" Shortly thereafter, he refined his proposition into a less subjective question and asked whether machines can mimic interactions that they have been exposed to (via data from the interactions that are fed to the machine). In his original example, he asked whether a machine could mimic the interactions of three agents, a man (A), a woman (B) and an interrogator (C). The role of the interrogator (C) was to interrogate data from A and B, and based on the data make a judgement as to the gender of the agent.

A number of obstacles to artificial intelligence and robotic process automation presented themselves around this time, most notably, the exorbitant cost of computers and a lack of ability to store large amounts of data. Although the ability to store results on a computer had developed by 1950, this technology was in its infancy and access to computing machinery was limited to large, well-funded institutions as a result of the expensive costs associated with the technology.

In 1956, Cliff Shaw, Allen Newell and Herbert Simon further explored this field. Their programme, the Logic Theorist, was the first computer programme that sought to mimic the problem solving ability of humans.[207] Many consider this programme to be the first piece of artificial intelligence software ever created. The programme was used to prove 38 of the first 52 theorems in the *Principia Mathematica*, one of which resulted in a more refined solution than the original authors had produced. These discoveries were presented at the Dartmouth Summer Research Project on Artificial Intelligence in 1956, and paved the way for many future researchers to further explore the field of artificial intelligence.

In the years that followed, many ideas were hampered by the infrastructure required to produce artificial intelligence as the cost and power of computers were not able to support this level of research. Over time, this started to change quite rapidly as the costs came down while the capacity and speed of computers began to increase at an exponential rate. The rise of the personal computer in the early 1980s provided a significant departure from previous technological advances in computers, as these were intended for use by individuals rather than as a mainframe, which had previously been the case. As computers became more accessible, individuals had more bearing on the pursuit of inventing artificial intelligence than ever before. As the power of, and access to, computers continued to improve, the real power of computers continued to be harnessed and enhanced. In 1997, a computer named Deep Blue, which was developed by IBM, beat the World Chess Champion, Gary Kasparov, in a game of chess.[208] A similar feat took place in more recent times when Google's AlphaGo machine beat Ke Jie (the Go World Champion) at the board game Go in 2017.[209]

Compared to other artificial intelligence discoveries, these may seem to be somewhat trivial as they involve games, yet the design of a programme that can not only mimic but exceed the level of thought of even the most skilled humans cannot be overlooked. As more research is done and these discoveries are applied to more far reaching uses, artificial intelligence can present a serious threat to the very humans who invented it. We already live in a world where a number of process driven tasks are being performed by machines rather than humans. Imagine an assembly plant 100 years ago versus today. One hundred years ago, the assembly line would have been characterised by a large number of humans each doing their part as part of a larger process, which ultimately resulted in finished products being available to the consumer. In the modern-day era, the number of humans required to assemble the same item is greatly reduced as a result of machine learning and robotics.

For example, at BMW's production plant in South Carolina in the USA, the X5 and X6 models are produced almost exclusively using robots, with only a skeleton crew of humans ensuring that the robots are operating correctly and without any malfunctions. This is in contrast to the way cars were made, for example, in Chevrolet's manufacturing assembly line in 1936, when the process was highly labour intensive. This assembly line employed thousands of employees and humans did the majority of the work, even employing toolmakers and other artisans to hand-make many components of the vehicles.

WHAT IS ROBOTIC PROCESS AUTOMATION AND HOW DOES IT DIFFER FROM ARTIFICIAL INTELLIGENCE?

The concepts of robotic process automation and artificial intelligence are often used interchangeably when the general public discusses these concepts. The origins of the two concepts share the same roots, however there are fundamental differences between them. Robotic process automation can be described as software that allows a machine to perform actions formerly performed by humans by following a strict set of rules which can be replicated.[210] In other words, robotic process automation involves creating software that allows machines to perform rule bound processes. Within a motor vehicle assembly plant, a machine may be coded in such a way that it knows upon receiving the body of the vehicle that the doors should be added before allowing the vehicle to move to the next station. The rules of this process are clear and allow zero scope for interpretation or self-learning, as the role of the machine is based on instruction received from the robotic process automation software.

The benefit of this type of software lies in its efficiency from a cost and production point of view. The accuracy and speed of a well coded robot is superior to that of a human. Unlike a human, a robot does not require a salary to perform its task, merely a capital outlay upfront and maintenance costs. Robots can also operate 24 hours a day, seven days a week; do not get sick (although they do require repairs occasionally); do not get disgruntled; and do not negotiate an annual salary increase each year. A robot's level of precision is also unmatched by humans, as robots can replicate the same process over and over without any deviation from the parameters which it has been set. The disadvantages associated with robots in the workplace are mostly surrounding their impact on the employment of individuals. Reverting to the assembly line example, it is not uncommon for factory floors that were once full of employees to now only contain a fraction of the staff headcount and a number of machines. The benefits available from a business point of view are enormous and are easily understood; more uninterrupted production at a lower cost. In the context of the South African economy, however, there are particularly harsh disadvantages as South Africa's unemployment rate ranged between 23% and 27% between 2008 and 2018. This unemployment rate is high by global standards and is a particular hindrance to economic growth in South Africa's labour abundant economy.

Artificial intelligence goes a step further than robotic process automation and allows the machine to interpret information and even self-learn based on the information received and the outcomes of various events.[211] Deep Blue (IBM's chess playing computer) and AlphaGo (Google's Go playing computer) computers are examples of artificial intelligence as these computers analysed the current state of the game before selecting their next move. These machines also learned from their mistakes as they played their respective games iteratively, each time learning from their mistakes and improving their likelihood of a positive outcome (winning the game). Artificial intelligence is designed to mimic a human's ability to learn, as opposed to robotic process automation which is designed to allow a machine to perform a predefined process. Robotic process automation requires structured information as it does not possess the interpretive skills required to make judgements. Artificial intelligence, on the other hand, can make use of semi-structured information as the software; it has the ability to make judgements on where the information is, even if it is not always in the same place.

If a piece of robotics process automation software was given a purchase order to process, it would require the purchase order number to always be in the same location (this software would need a data input template). Conversely, artificial intelligence software could locate the purchase order number on the form, even if it has moved from the top left to the bottom left hand corner of the purchase order.

The key differentiator between these software types is that robotic process automation software operates **within** "the box", while artificial intelligence can operate **outside** "the box".

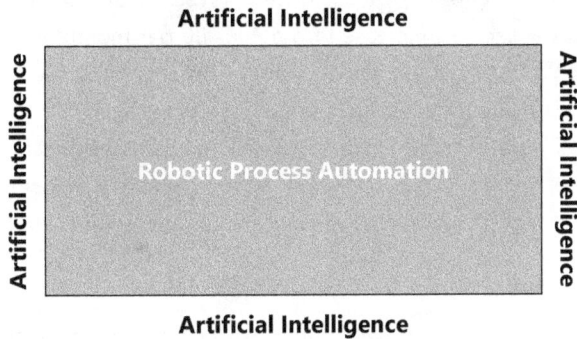

Figure 10.1: Robotic process automation vs. artificial intelligence

HOW ARE ROBOTIC PROCESS AUTOMATION AND ARTIFICIAL INTELLIGENCE BEING USED IN MODERN DAY SOCIETY?

In describing the differences between the two technologies, we have briefly touched on how these technologies can, and have, been applied in modern day society. Both technologies seek to improve efficiency, reduce errors, reduce costs and limit wastage, but they have different futures as technology within these fields continues to develop. This section provides a case study of how robotic process automation and artificial intelligence have been applied in the modern era.

CASE STUDY **Robotic process automation**

Telefonica O2 (O2) is the second-largest mobile telecommunications provider in the United Kingdom. As of 2015, O2 had 24 million customers, operated 450 retail stores and employed 21,580 employees.[212] O2 faced the challenge of scaling up its "back office" operations (such as entering data into its CRM system) while keeping costs low to remain competitive in the market. In 2004, O2 moved its "back office" operations from the UK to India as part of their business process outsourcing and employed 200 employees in India and 98 employees in the UK. By 2009, the employee head count grew to 375 in India and was reduced to 50 employees in the UK, but costs were increasing as a result of rising wages in India and the company was reaching the apex of the benefit that could be enjoyed by outsourcing. These processes and the number of transactions increased from 400,000 per month to more than a million per month, but the cost per transaction was rising. This prompted the Head of Back Office services to roll out his vision to reduce the number of employees by 50%, reduce the average response times by 50%, and reduce back office failure customer calls by 50%. In 2010, O2 managed over 60 processes, which consisted of approximately 400 sub-processes. These processes required a significant revamp if the vision of the Head of Back Office was to be realised.

In 2010, O2 decided to implement two robotic automation processes as pilots for similar projects in the future. These two processes involved high volume, low complexity processes. The processes were performing SIM swaps and the application of a pre-calculated credit to a customer's account. The pilot was completed within two weeks and the results were positive. O2 cast their net wider and sought other areas where they could improve using robotic process automation. They performed a case study to compare how the results of ten processes would differ when using robotic process automation versus business process management systems. The robotic process automation processes would pay back their costs within ten months and would result in a financial benefit of nearly £1 million per year. In contrast, the business process management system would only break even after three years. The core reason for the difference in the costs was that the business process management system required more staff such as developers and SCRUM teams, unlike the robotic process automation system. Ultimately, O2 automated fifteen processes using just three staff and robots, which accounted for 35% of all back office transactions by the end of the first quarter of 2015.

CASE STUDY

Artificial intelligence

Under Armour, an American sportswear and apparel manufacturer, made use of IBM's Watson (IBM's AI platform for businesses) to create its Under Armour Record™ application.[213] This application is a "cognitive coaching system", which was designed to serve as a personal health assistant that provides users with real-time, data-based coaching based on sensor and input data for sleep, fitness, activity and nutrition. This application also makes use of data which reflects the environment within which the athlete is training to determine if there are any factors affecting their training data.

The application allows the user to track their physical performance against their goals by linking data sourced from their Under Armour Band which tracks their activity. The "just like me" feature allows the user to compare themselves against other users using the same application, provides insights into the user's performance relative to others, and makes suggestions based on the data collected. The user can also track their performance against their friends or athletes who make use of the application.

The Under Armour Record™ application was very positively received by users, and in 2016 it led to a 51% increase in the fitness related accessories which integrate with this application.[214]

Artificial intelligence

A second, more advanced case study of how artificial intelligence has been implemented is Google Assistant. Google Assistant is what it sounds like, a virtual assistant that is similar to Apple's Siri and Amazon's Alexa. It allows users to engage with it via their speaking voice or by typing, and is primarily used for tasks such as searching the internet, adjusting phone settings, updating calendars, setting memos, linking to smart automation devices etc.

At Google I/O 2018, an annual event which attracts developers and technologists from around the world to talk about the future of technology, Google CEO Sundar Pichai demonstrated how Google Assistant booked a haircut appointment on behalf of its user.[215] The nature of the call was so seamless that it was hard to believe that this was a computer, acting on behalf of a human, making an appointment with a human. Below is a log of how the call went:

The user's instruction to Google Assistant (GA) was: "Make me a haircut appointment on Tuesday morning anytime between 10 and 12."

Hair Salon: "Hello how can I help you?"
GA: "Hi, I'm calling to make a women's haircut appointment for a client. I'm looking for something on May 3rd."
Hair Salon: "Sure, give me one second."
GA: "mm-hmm."
Hair Salon: "Sure, what time are you looking for around?"
GA: "At 12pm."
Hair Salon: "We do not have a 12pm available. The closest we have to that is a 1:15."
GA: "Do you have anything between 10am and 12pm?"

Hair Salon: "Depending on what service she would like. What service is she looking for?"

GA: "Just a women's haircut for now."

Hair Salon: "Okay, we have a 10 o'clock."

GA: "10am is fine."

Hair Salon: "Okay, what's her first name?"

GA: "The first name is Lisa."

Hair Salon: "Okay, perfect. So I will see Lisa at 10 o'clock on May 3rd."

GA: "Okay great, thanks."

Hair Salon: "Great, have a great day. Bye."

END CALL

At the conclusion of this demonstration, the audience erupted into applause, clearly impressed by the demonstration of the advanced artificial intelligence which had been displayed to them. Mr Pichai spoke about the call and said that it was the culmination of many years of work and drew on many different technologies that they had been working on. One of the most impressive features of the demonstration (and there were many) was how the Assistant sounded like a human; it was very difficult to distinguish that this was in fact a robot and not a person making the phone call on their client's behalf. This demonstration left the audience thinking about the future and what this kind of technology will evolve into in the coming years.

WHERE IS ROBOTIC PROCESS AUTOMATION AND ARTIFICIAL INTELLIGENCE HEADING AND HOW CONCERNED SHOULD WE BE?

People can be resistant to change, particularly when it makes them uncomfortable. The pace at which technology is advancing has been increasing and will continue to do so in the future. Naturally, there are those who will embrace change and those who will resist it. Those who embrace change see the potential of technology, while those who resist it are fearful of its effect on society. The future of artificial intelligence is bright and filled with many possibilities. *The Wall Street Journal*, *Forbes* and *Fortune* all called 2017, "The Year of AI". Going forward it can be expected that the evolution of artificial intelligence will continue at a rapid rate. The following list of the main expected changes through AI and automation are listed below:

- Just as in previous industrial waves of change, AI will enable the transformation of communications, energy and transport.

- It will continue to change manufacturing roles but this technology will also impact service, professional and leadership roles.
- Current human work will be delivered by very powerful machines.
- Developed applications will be very quickly rolled out globally.
- There will be knock-on impacts – a change in a key technology, such as automated cars, has a major impact on many related industries.
- There will be global benefits from improved medical, education and farming innovations.
- AI innovations are closely linked to other transformational changes taking place today such as renewable energy, electric cars and digital communications.

Listed below are a few quotes from prominent businesspeople regarding where they believe artificial intelligence is heading in the next few years.[216]

Artificial intelligence in engineering

"2018 will be the year deep learning starts a revolution in engineering simulation and design. Over the next three to five years, deep learning will accelerate product development from years to months and weeks to days to create a new paradigm of rapid innovation in product features, performance and cost." (Marc Edgar, senior information scientist, GE Research)

Artificial intelligence in medicine

"2018 will be the year AI becomes real for medicine. We're going to move from algorithms to products and think more about integration and validation, so that these solutions can move from concepts to real, tangible solutions for our doctors. By the end of next year, I think around half of leading healthcare systems will have adopted some form of AI within their diagnostic groups. And while a lot of this adoption will happen first in the diagnostic medical specialties, we're seeing solutions for population health; hospital operations and a broad set of clinical specialties quickly follow behind. In 2018, we'll begin the adoption of a technology that may truly transform the way providers work, and the way patients experience healthcare, on a global scale." (Mark Michalski, executive director, Massachusetts General Hospital and Brigham and Women's Centre for Clinical Data Science)

Artificial intelligence in technology

"AI is going to affect 25 percent of technology spend going forward. The key topic is how organizations and the human workforce will cope with the changes that AI technologies will bring." (Nicola Morini Bianzino, managing director of Artificial Intelligence and Growth & Strategy Lead of Technology, Accenture)

Artificial intelligence and biometrics

"Thanks to AI, the face will be the new credit card, the new driver's license and the new barcode. Facial recognition is already completely transforming security with biometric capabilities being adopted, and seeing how tech and retail are merging, like Amazon is with Whole Foods, I can see a near future where people will no longer need to stand in line at the store." (Georges Nahon, CEO, Orange Silicon Valley; President, Orange Institute, a global research co-laboratory)

Artificial intelligence and daily life

"Robots are going to get better at complex tasks that humans still take for granted, like walking around a room and over objects. They'll get better at mastering boring, normal things. I'm looking forward to seeing progress in NLP tasks as well, since right now we've got a ways to go. We're going to see more and more products that contain some form of AI enter our lives. Waymo's level 4 autonomous vehicles are deployed on the road now. So all this stuff that's been tested in the lab will become more common and available. It will touch more lives." (Chris Nicholson, CEO and co-founder, Skymind.io)

As the span of uses of artificial intelligence and robotic process automation increase, machines are performing more and more tasks which have traditionally been performed by humans. As a result of the improved efficiency and lower costs that are attainable using technology, some humans are being replaced by machines, which has created a wave of concern in societies around the world. This begs the question: "When does technology stop helping and start harming society?" This is a question of ethics and responsibility, as uncapped development and implementation could lead to large scale job losses and increase income inequality around the world.

However, it is not all doom and gloom in our daily lives. Whilst some jobs will decline with automation and AI, others will increase. Driving, manufacturing, data analysis and routine tasks will be replaced, whilst jobs in renewable energy, the application of AI systems, caring services, entertainment and vacations will be in demand.

The insights provided by the captains of industry quoted all point to the same outcome; artificial intelligence is being embraced and used in a multitude of fields for numerous purposes and the span of uses is increasing exponentially.

Finding a balance between technological advancement, efficiency, cost and human jobs is vital to society's future. How this balance should be found is a question which is easier asked than answered. Below are a few key areas to address when discussing issues of artificial intelligence and how they affect human lives.

Employment

As far back as the industrial revolution, machines and humans have had an uneasy relationship in the workplace. Machines may be useful to humans in a factory, for example, as they speed up production and reduce waste, but as more machines are employed fewer humans are needed than before. As the pace of technological innovation continues to increase, we are seeing more and more traditional jobs being lost as a result of automation. Ethically, we should ask ourselves about the purpose of machines. Are they meant to replace or enhance people's jobs?

Why is it important for Human Resources as a field to learn to adapt to AI and automation?

- It is a transformational change happening now.
- Many organisations will have their traditional business models challenged and have to change.
- It will impact and change virtually every role in the next 20 years, changing the way work is done and in turn changing societies.
- HR will change but HR will also need to be at the forefront of the wider organisational change.

An example of an HR process that has changed dramatically due to AI and automation is recruitment. Recently, recruitment, a task traditionally performed by humans, has been performed by machines in some cases. Recruitment can be a slow and tedious process as the recruiter screens multiple CVs searching for the right individual. As a result of this slow process, machines have been introduced which can replace the human and select appropriate candidates in a fraction of the time. Should these machines be used to replace the human who was employed to do this job or to assist the human to speed up the process?

Economic inequality

In 2017, it was stated by Oxfam that the world's richest eight people own the same amount of wealth as the poorest 3.6 billion people in the world (50% of the global population). This is an incredibly damning statistic and is evidence of just how unequal global society is. Historically, those who have controlled the resources of production have held far more wealth than those who sell their labour to the labour market (those who have jobs), and those who sell their labour enjoy more wealth than those who do not have jobs. If machines replace people, the number of jobs will decline, increasing the number of people who do not have jobs. The handful of individuals who own the factors of production (the machines) will benefit from the lower costs and more efficient production processes at the expense of those who have lost their jobs. The net result will be the rich getting richer while the poor get poorer, and wealth inequality will continue to grow.[217]

Security and crime

> "With great power comes great responsibility." (Uncle Ben – Spiderman's uncle)

The source of the quote may be somewhat comical, but the message rings true. As artificial intelligence advances and gets used for a number of positive purposes, the potential of it being used for immoral purposes increases as well. Society needs to guard against and be protected from those who would use technological evolution for crime rather than positive purposes.

Technology and our daily lives

Twenty years ago, if you wanted to contact your friend you either had to pick up your home telephone and phone their house, or drive over to their house and hope that they were at home. Since the invention and widespread use of the cellular phone, contacting friends and family has become a far easier task. The emergence of various forms of social media and video calls (as examples) has made people more contactable and available than ever before. Referring back to the Google Assistant example, soon a person will not even need to personally use a phone, as a machine can make an appointment on their behalf. Although this technological advancement has been positive for numerous reasons, it also has its negative social side effects, both of which have affected the way humans live.

Artificial intelligence and rights

As artificial intelligence provides robots with the means to mimic human behaviour, how are the issues of legal responsibility and rights to be handled? A humanoid robot named Sophia, developed by Hanson Robotics, mimics human appearance and behaviour to such an extent that it became the first robot to be granted citizenship of a country (Saudi Arabia). This unprecedented move can be considered a simple publicity stunt, yet what are the consequences of this move? Should this citizenship confer rights and responsibilities onto the robot? Can it enter into a legal contract? Should it pay taxes? Can a robot be held liable for committing a crime? These are some of the questions which must be asked should robots move from possessions to having rights and responsibilities.

CONCLUSION

Artificial intelligence and robotic process automation began as an idea which has been translated into reality at an ever increasing rate. Initially, the idea of this kind of technological advancement was marred by a lack of technology to support this development, but as computers became cheaper and more powerful, the means to create the envisaged reality was created. As technology develops further, the role of technology in society must be considered. Machines have the potential to replace humans within a number of traditional jobs, but should technological advancements be used to replace people in the interest of profits and efficiency, or augment people's roles without replacing them? The temptation to pursue profits within our capitalistic society is ever present, and has resulted in the wealth gap between the rich and the poor increasing steadily over time. South Africa, which has the worst distribution of income in the world (according to the World Bank), is particularly aware of the societal problems caused by high levels of unemployment and a skewed distribution of wealth.

If humans and machines are to co-exist in a manner which is beneficial to society as a whole, the role of human beings in the workforce must be preserved and maintained. This does not mean that there is no role for technological advancement and machines, but they should assist humans rather than replace them. As the implementation of machines in the workplace occurs, it is important to redeploy human resources into other areas of the business. Artificial intelligence and robotic process automation, if used responsibly, provide an opportunity to reduce the cost of living, as costly, inefficient processes are replaced with cheaper, more precise processes. However, the benefits of the reduced cost and increased efficiency can only be enjoyed if consumers are able to purchase the goods and services that have been produced. Businesses and organisations need to adopt the profit, people and planet ethos to develop strategies that improve society through AI and automation instead of reducing living conditions. It is the role of society to ensure that we create an ethical framework which allows humans to enjoy the benefits of artificial intelligence and robotic process automation, but these benefits should not be realised at the expense of human jobs.

CHAPTER 11

THE ROLE OF HUMAN RESOURCES IN THE NEW WORLD OF WORK

INTRODUCTION

Agile methodologies originated in the IT space, but it has become clear that creating a new organisation within an old organisation is not optimal. As IT practices have changed and new ways of work have started to become embedded, the necessity for an enabling environment has become apparent. The interface between Agile and traditional/waterfall practices causes conflict and frustration. Support functions, like Human Resources, have been challenged to evolve and adapt their practices to enable new ways of work, and to be honest, it is about time. The vast majority of HR functions still follow the David Ulrich model, which is now 21 years old.[218] It is time for a change and HR is ideally positioned to lead this change. Successful agility is dependent on a shift in culture and values – becoming more transparent, collaborative, user-centric and team-based. HR practices can be used to both drive and shape Agile transformation.

TRADITIONAL HR – TIME FOR CHANGE

The term 'human resource' was coined by economist John R. Commons in his book, *The Distribution of Wealth,* which was published in 1893.[219] Before this time, the first form of human resource management, Industrial Welfare, was embodied in the 1833 Factories Act, which stated that there should be male factory inspectors. In 1878 legislation was passed to regulate the hours of work for children and women by having a 60 hour week.[220] The main purpose of early HR was to address misunderstandings between employees and their employers.

By the 1920s the field of HR became known as 'personnel administration', and focused on the technical aspects of hiring, evaluating, training, and compensating employees. The advent of globalisation, deregulation and rapid technological change in the 1970s led to the emergence of what we know and recognise today as the discipline of Human Resource Management.[221]

A brief history of modern HR

"After World War II, when manufacturing dominated the industrial landscape, planning was at the heart of human resources: Companies recruited lifers, gave them rotational assignments to support their development, groomed them years in advance to take on bigger and bigger roles, and tied their raises directly to each incremental move up the ladder. The bureaucracy was the point.

"By the 1990s, as business became less predictable and companies needed to acquire new skills fast, that traditional approach began to bend — but it didn't quite break... For the most part, though, the old model persisted. Like other functions, HR was still built around the long term. Workforce and succession planning carried on, even though changes in the economy and in the business often rendered those plans irrelevant. Annual appraisals continued, despite almost universal dissatisfaction with them.

"Now... rapid innovation has become a strategic imperative for most companies, not just a subset. To get it, businesses have looked to Silicon Valley and to software companies in particular, emulating their Agile practices for managing projects."[222]

Despite the enormous change that the past two decades have brought to the world and the world of work specifically, the majority of organisations continue to use traditional HR models and practices. In 1997, David Ulrich published his seminal book, *Human Resource Champions: The Next Agenda for Adding Value and Delivering Results*, and the 'Ulrich Model of HR' emerged. The ultimate goal of the model was to shift the role of HR from administration to strategy.[223]

While a great deal of time and money has been poured into creating the shared service transactional environments that Ulrich called for, many organisations have not managed to move their HR business partner communities into the transformational and strategic realm. While there are often structural impediments to implementing transformational HR practices (like multi-layered, hierarchical structures characterised by bureaucracy and stifling risk and control systems), the biggest impediments are culture and leadership. The shifts that are required to move from a tactical to a more strategic HR function are tabulated below in Table 11.1.

Table 11.1: The shifts required to move HR from a tactical to strategic orientation (Adapted from Denning)[224]

Tactical/Transactional HR	Transformational/Strategic HR
From:	To:
Inside-out thinking (The focus is on internal processes)	Outside-in thinking (The focus is on the customer)
Managers as controllers	Managers as enablers
Coordinating work through bureaucratic processes	Coordinating work through dynamic linking
Economic values as a driver	Values as a driver
Top-down command	Adult-to-adult conversations
Outputs	Outcomes

In many traditional organisations, HR is often seen as an impediment to agility as opposed to an enabler. People practices have not kept pace with the demand for a digital, seamless employee experience that can be customised. If an organisation is invested in adopting Agile practices, HR and its people practices will need to evolve... and quickly.

CHANGING PEOPLE PRACTICES

Some people practices have started to evolve and align with the demands of an Agile organisation. The most notable and visible shifts can be seen in performance management. In a 2017 Deloitte survey, 79% of global executives rated Agile performance management as a high organisational priority[225], but other HR processes are also starting to change. Many of these practices are discussed in detail in the chapters of this book, but here is a brief summary of the primary changes.

Performance management

The unit of production has changed from the individual to the team, and performance management has shifted to enable a team-based approach. Team goals and key performance indicators (KPIs) replace individual goals and KPIs. Ongoing feedback is a critical lever of success and multi-directional feedback becomes key to driving performance improvement. Performance reviews are held more regularly and often dove-tail with quarterly business reviews.

Recruitment and on-boarding

In the new world of work, scarce skills are highly sought-after and lengthy recruitment and on-boarding practices can dissuade talent from joining your organisation, thus it is important to keep these to a minimum. Finding good candidates is hard enough; losing them to competitors while you are busy trying to schedule that fifth interview can make the sourcing of talent very difficult.[226]

Recruitment practices are becoming more innovative to attract both active and passive talent. For example, hackathons have become recognised as a great way to find IT talent. These are events where computer programmers and others involved in software development meet, suggest ideas, form teams, and then build solutions. Hackathons are often competitive and offer prizes to drive participation.

Social media, professional meetings and meet-ups should all be used to source talent. You cannot wait for candidates to come to you – you need to go to them. Recruitment in an Agile organisation becomes everybody's job and is not only an HR practice. The use of technology and artificial intelligence is increasingly being used to source and screen candidates, reducing the time it takes to source and improve candidate selection.

Once you have recruited a candidate, getting them up and running as quickly as possible supports both productivity and a positive employee experience. Induction/orientation is critical to make sure that new hires are exposed to the organisation's strategy, structure and ways of work. Get the basics right – make sure that employees have access cards, company email addresses, a computer, parking etc. on day 1 if possible.

🔍 **CASE STUDY**   **Recruitment and internal mobility (ING)²²⁷**

Case study – Recruitment practices pertain to internal mobility as well as the external sourcing of new candidates. Organisations that decide to embark on a full-scale Agile transformation need to consider their existing workforce and whether their employees are fit for the new organisation in terms of culture, skill-set and existing role/placement. Bart Schlatmann, the CIO of the Dutch banking group, ING, recalls when every single person in the company was obligated to re-apply for their job:

> "I still remember January of 2015 when we announced that all employees at headquarters were put on "mobility," effectively meaning they were without a job. We requested everyone to reapply for a position in the new organization. This selection process was intense, with a higher weighting for culture and mind-sets than knowledge or experience. We chose each of the 2,500 employees in our organization as it is today—and nearly 40 percent are in a different position to the job they were in previously. Of course, we lost a lot of people who had good knowledge but lacked the right mind-set; but knowledge can be easily regained if people have the intrinsic capability."

Learning and development

Agile organisations both thrive and rely on the adoption of a learning culture. As technology and innovation disrupt traditional skills and requirements for success, it is critical to enable employees to become life-long learners and provide access to ongoing learning and development (L&D). While there is no one specific learning type that works best in an Agile setting, there are certainly some trends which guide Agile L&D strategies:

- The use of, and reliance on, traditional classroom-based training is declining.
- Where classroom-based training is used, the shift is to move toward more experiential learning laboratories where participants learn by doing, and not just by listening to a facilitator/lecturer.
- Nano-learning is becoming highly desirable. This relates to bite-sized pieces of learning where employees can take a short amount of time to learn something new, useful and relevant while they work.

- Accessibility to learning needs to be 'always-on' and flexible. Many employees choose to learn outside of working hours and want access to learning materials on mobile phones, personal devices and work devices.
- There is so much L&D content available that it can become overwhelming and difficult to determine which content is credible. Organisations need to curate as well as create their own content to help employees navigate the depth and breadth of content available.
- Promoting the development of cross-functional skills is key, so employees should be able to learn about a wide range of topics and not be confined to areas of study that only pertain to their current job or organisation.
- Learning agility is becoming a key competency to influence recruitment and promotional decisions. It is no longer possible or necessary to have all the knowledge available, however it is necessary to have a curiosity and desire to learn and adapt to changes in your work and the wider world of work.

Recognition and reward

Just like performance management, recognition and reward practices need to evolve to cater for the team and not just the individual. In both recognition and reward, the trend is to move toward more regular incentives. The incentive should be given as close to the event as possible, and annual bonuses and recognition awards are no longer seen as fit-for-purpose. HR needs to provide a simple framework in which salaries and bonuses are competitive and equitable, and recognition decisions can be taken by the team.[228]

Transparency and fairness are essential components to consider when designing team recognition and reward practices. The purpose of team-based reward and recognition is to drive not only high performance, but also the right behaviours. Rewards can be divisive and lead to the breakdown of team effectiveness if not managed carefully and fairly.

Rewards extend beyond money/pay, and there are a number of ways that you can incentivise team performance without using compensation. Time off, access to learning and development, exposure to different work, selection for innovative projects and public appreciation should all be considered as valuable recognition and reward tools.

Organisation design

Agile methodologies call for different organisational structures. Whereas traditional organisations are built around hierarchies, rigid structures and product-centric processes and practices, Agile organisations have much flatter structures, networks of teams and client-centric processes and practices.

Aghina, De Smet, Lackey, Lurie and Murarka from McKinsey described five organisation design trademarks of Agile organisations:[229]

	Trademark		Organizational-agility practices
Strategy	North Star embodied across the organization		• Shared purpose and vision • Sensing and seizing opportunities • Flexible resource allocation • Actionable strategic guidance
Structure	Network of empowered teams		• Clear, flat structure • Clear accountable roles • Hands-on governance • Robust communities of practice • Active partnerships and ecosystem • Open physical and virtual environment • Fit-for-purpose accountable cells
Process	Rapid decision and learning cycles		• Rapid iteration and experimentation • Standardized ways of working • Performance orientation • Information transparency • Continuous learning • Action-oriented decision making
People	Dynamic people model that ignites passion		• Cohesive community • Shared and servant leadership • Entrepreneurial drive • Role mobility
Technology	Next-generation enabling technology		• Evolving technology architecture, systems, and tools • Next-generation technology development and delivery practices

Figure 11.1: Five organisation design trademarks of Agile organisations

These trademarks enable organisations to balance stability and dynamism and thrive in the move toward organisational Agility.

CASE STUDY

Redesigning a global HR function to support Agile transformation (BBVA)

Banco Bilbao Vizcaya Argentaria, S.A. is a multinational Spanish banking group. It has a presence in over 30 countries, provides services to 75 million customers, and employs almost 132,000 staff members. BBVA began their HR transformation journey at the end of 2016 by creating a pool of people fully dedicated to projects in all Talent & Culture (T&C) units across countries, moving 10% of the team to this new project-based organisation. In parallel, they started to analyse the opportunity to create shared-services centres in every country. Within

149

a few months it became clear that having this dual organisation (the old organisation and the emerging Agile structure) was not very effective. By mid-2017, the head of T&C in Spain decided to move forward by transforming the whole team into a fully Agile organisation. Over the next nine months they were able to go through the same process in every country, and today they have a T&C team of more than 2,000 people in 10 countries working under a fully Agile organisation and governance model.

BBVA has broken up the previous functional units and hierarchies and reorganised the team along four different groups:

1º Front
This team of business partners offers strategic advice and support to internal customers: areas, managers and employees. Business partners have to play a strategic and proactive role in giving service to internal areas, based on a very good knowledge of their needs and priorities. They also have to act as coaches for managers and as a point of contact for employees through their life cycle in the organisation. They typically represent 10-15% of the team.

2º Disciplines
These expert teams have the role of defining the strategy and developing the models, policies, tools and platforms for their respective areas of expertise (such as talent management, compensation and benefits, internal communication, and organisation design). They ensure the connectivity of people in execution teams through global communities of practice, in which practitioners in their field share knowledge and best practices, and even co-create new models and platforms. Discipline teams are typically senior but very small (just 2-3 people by discipline), representing no more than 10-15% of the total team.

3º Solutions development
This pool of professionals is fully dedicated to executing projects or building new solutions following Agile scrum principles. They constitute multi-disciplinary teams with the autonomy to organise their work and end-to-end accountability and capacity to execute. These scrum teams typically work in 2-3 week sprints following an iterative, incremental process to continuously learn from (internal) customer feedback. They are dynamically assigned, on a quarterly basis, to the evolving strategic priorities of the area through a staffing process. Ideally they should represent at least 25-30% of the team.

4º Employee experience
These groups of teams are empowered to execute all end-to-end processes in the function and deliver value to internal customers using Kanban. They have a big impact on employee experience, operational excellence and productivity. By concentrating all the processes that were previously fragmented into different units, there is a clear opportunity to stop doing things that do not add significant value; apply process engineering to redesign processes for better quality and efficiency; and apply automation, robotics and machine learning techniques to raise productivity. Furthermore, they define a catalogue of services to be provided to internal customers, linked to specific KPIs and service level agreements, so they can measure and continuously improve quality of service. These teams normally represent 40-50% of the area.

> This new organisational design constitutes the backbone for a new governance model demanding new roles and responsibilities, new team ceremonies, new people management models (such as project staffing and mentor leaders for the Solutions Development team), new communication tools to increase openness and transparency, etc.
>
> The ultimate goal of the Agile transformation was to promote a cultural transformation that puts execution teams at the centre of the organisation and transforms managers into servant leaders. Leaders now focus on giving strategic guidance to the teams, helping them to solve impediments and acting as coaches to help everyone build new skills and mind-sets. Teams benefit from increased visibility, empowerment and accountability.[230]

Organisational development

Agile transformations are essentially people transformations, underpinned by culture and leadership shifts. The implementation of Agile practices and structures in the absence of a change in mind-set and behaviour is unlikely to lead to any substantive gains. Introducing Agile ceremonies is easy; placing employees in small self-managing teams is also a simple task. The complexity lies in the relinquishment of traditional command and control cultures and creating an environment where psychological safety flourishes.

One of the biggest challenges is helping leaders to move from controlling the work to enabling the work. This requires leaders to lead using influence and not positional power.

Another fundamental challenge is enhancing team effectiveness. The construct of collaboration becomes a defining feature of the Agile organisation's structure and culture. Collaboration, breaking down traditional silos, and reducing hierarchy, bureaucracy and unhealthy internal competition need to become key focus areas of organisational development practices.

THE EMERGENCE OF EMPLOYEE EXPERIENCE AND EMPLOYEE-CENTRED DESIGN

In the past, HR rarely consulted employees on the design of people policies, practices and processes. Design principles like standardisation, cost-efficiency and market-related benchmarking were used to develop "best practices". In Agile organisations, it is increasingly clear that these policies, practices and processes do not serve employees or organisations well. An organisation can offer employees good benefits, fair pay and meaningful work, but if they do not pay attention to the employee experience, the attraction and retention of talent will become a problem.

Employee experience is how it FEELS to work at an organisation; it is a combination of tangibles (physical workspace, pay, benefits, performance management practices, the quality of the food and coffee available, etc.) and intangibles (treatment by colleagues and leaders, perceived fairness and transparency, the ease of getting things done, etc). One of the roles of HR is to provide a seamless employee experience that makes working for an organisation easier. Employees and job seekers today are super-connected, impatient for quick results and are placing pressure on organisations to become more tech-savvy. HR needs to put systems in place that allow for the automation of routine paperwork, the handling of growing volumes of data, the reduction of administration for employees and line managers, and the availability of intuitive self-service transactional services and systems.[231] Taking leave, being paid on time, managing benefits and using performance and talent management systems should be as effortless as possible. In a sense, HR should be 'invisible' in its execution of transactional people practices.

When these people practices require refinement or a complete overhaul, Agile HR functions turn to their employees for assistance. Just as customer-centred design and design thinking methodologies are being used in business, employee-centred design is making its way to the forefront of new people practices design. Employees should help to co-create the practices that they will use. When they do this, not only is the resulting practice better adopted, but the employee experience is enhanced. Old fashioned notions of paternalism in the workplace are being challenged by employee-centred design and this provides employees with a sense of agency.

Most of the new Agile people practices rely on agency. For example, employees are urged to own their learning and development journeys, to initiate ongoing performance management feedback sessions, and to ensure their own skills are future-fit. If we want employees to drive and build their own employee experiences, we need to enable them to contribute to the design of the practices that underlie this experience.

Employee experience is often used synonymously with the term 'Employee Value Proposition' (EVP). Although the components of the two may seem similar, there is a significant difference between them; an EVP is the rational and intellectual understanding of the people deal – what your organisation provides to employees and what they expect in return. An EVP can be articulated on paper and is often used by recruiters to 'sell' the organisation to candidate employees. Employee experience, on the other hand, is more emotive; it is how an employee experiences the overall organisation and often cannot be clearly articulated. There is no 'ideal' or 'one-size-fits-all' employee experience — what makes one employee happy may not matter much to another.

For example, some employees value the provision of free food and drinks and a physical workspace that is modern and attractive. To other employees, this makes no difference. They may say that the benefit of flexible work practices far outweighs the benefit of working in a great building. Some employees will be excited about internal mobility and see this as a way to grow and develop new skills. Other employees may see internal mobility as disruptive, prefer to specialise in one content area, and not become multi-skilled.

Employee experience is very subjective. The key to making employee experience work for your organisation is to allow employees customisation, choice and voice in determining what is important to them. It is true that you cannot have customised experiences, systems and practices for each employee, but mass customisation at least allows for employees to have some choice between fixed options.

THE VOICE OF THE EMPLOYEE AND SENTIMENT ANALYSIS

For employees to influence the employee experience and associated people practices, they need to have a voice within the organisation. The Voice of the Employee is a well-recognised construct that not only helps to create more employee-centred design practices, but also helps build psychological safety. Many organisations only provide employees with an opportunity to share their thoughts, ideas and opinions through structured surveys which are usually conducted annually. This may provide a channel and opportunity for feedback, but it falls short of creating a dialogue with employees and gauging employee sentiment.

Employees are, more than ever, challenging traditional ways of doing things, but four out of ten UK employees still do not feel it is safe to speak up, resulting in a large number of unheard voices and untapped ideas on how to address business challenges. An effective internal communications department does not equate to a 'listening strategy' — internal communications are most often just a vehicle for disseminating corporate messages. If these messages fail to resonate with employees' perceived reality and provide no room to communicate back up the 'chain of command', they will likely lead to cynicism and disengagement.[232]

Providing employees a voice helps to increase employee engagement, find out what what's working to engage employees and what they are thinking, and then find ways to replicate and extend these "best practices" throughout the organisation. Enthusiasm is a moving target and leaders need to continually monitor progress. Having a voice that is listened to matters to employees, and the best results are obtained when employees believe that their views are sought out, that they are listened to, and that their opinions count and make a difference. This leads to a greater likelihood of speaking out and challenging when appropriate.

Technology solutions have made the Voice of the Employee much easier to elicit, and also provide ways to gauge sentiment. A number of providers, such as Ultipro Perception and Willis Towers Watson, offer sophisticated platforms that use natural language processing and artificial intelligence to analyse and quantify employee voice and sentiment. Although quantitative surveys provide useful information, it is most often in qualitative feedback that the richness of context and employee voice is found. Before these technology platforms were available it would be too time consuming to elicit qualitative feedback from employees on a regular basis, as this resulted in employees (usually from HR) having to spend huge amounts of time reading the comments and conducting subjective thematic content analyses.

Natural language processing provides a solution to this, where employee comments can be themed and themes and sentiment can be quantified in real time. There is now no excuse for sticking to annual surveys and quantitatively heavy survey options. Agile organisations recognise the significant impact that the Voice of the Employee can make. When you constantly engage your employees it is possible to swiftly detect problems that are starting to brew and capitalise on the collective wisdom of the workforce.

The role of HR in addressing fear of Agile practices

When organisations start implementing Agile practices, it is not uncommon for employee fears and concerns to start to surface. These concerns include:

- feeling threatened, as some roles appear redundant;
- fear of breach of privacy due to the open work environment; and
- fear of loss of individual identity or the "heroism" factor that existed while working in silos.[233]

It is the role of HR to be a strategic partner, an employee advocate, and a mentor for change in these uncertain times.[234] HR is empowered to equip employees with the relationships, mind-sets, tools, and experiences to thrive in the complex transformation. HR needs to ensure that change management principles are embedded in how all people work, think and collaborate. Change can be unsettling, but it is also a great opportunity and can be exciting. HR professionals should act as change leaders and assist with the positive framing of the change.

Keeping employees motivated is difficult, especially during an Agile transformation which is characterised by uncertainty. HR needs to be creative in finding new ways to sustain motivated employees within the organisation. The focus should be on providing an environment of autonomy and developing mastery, which will encourage employees to approach their work with excitement.[235]

THE ROLE OF HR AS CHAMPIONS OF EMPLOYEE WELLBEING

There is no doubt that Agile ways of work result in high intensity work sprints, long hours and high levels of stress, thus there is a need for HR to advocate for a sense of balance and wellbeing at work. High performing teams are not placed under external pressure to work unrealistic hours at a fast pace; the pressure is internal and the desire to succeed and exceed expectations comes from the team itself. There is a danger that being overworked, stressed and miserable can become the norm for Agile teams.

Although speed to market may increase dramatically and results may be exceptional, HR needs to be the voice of reason and question the sustainability of the increased pace. HR should monitor absenteeism and be proactive about providing wellness resources such as mandatory vacation time, access to counselling services, and the availability of healthy food options in the canteen. Other aspects of wellbeing such as financial and emotional wellbeing should not be overlooked.

The notion of work-life balance is considered by many as outdated. People talk instead of work-life integration, where the distinction between the two is blurred. The culture of being "always-on" can lead to burnout, and while you cannot prescribe to employees how they work, HR needs to be cognisant that Agile ways of work can lead to heightened levels of stress and burnout, and should thus become the custodians of wellbeing at work.

CONCLUSION

As Agile methodologies and practices become more pervasive in organisations, HR has a responsibility and opportunity to redefine their role and enable future-fit people practices. This is an exciting time to be in HR as the field has been relatively stagnant for some time. The new world of work is calling for very different people practices, and being able to co-create these practices with business creates a platform for HR to take a far more strategic and transformational seat at the table. There is undoubtedly a role for HR in the organisation of the future, but whether that role is one that will be valued will be dependent on HR's appetite to reinvent their own purpose, mandate and ways of work.

ENDNOTES

Chapter 1 Endnotes

1   Bersin, J. (2017). *Predictions for 2017 - Everything is becoming digital.* Retrieved from https://www2. deloitte.com/content/dam/Deloitte/at/Documents/about-deloitte/predictions-for-2017-final.pdf

2   Merriam Webster Dictionary. (2018). *Definition of agile.* Retrieved from https://www.merriam-webster. com/dictionary/agile

3   Highsmith, J. (2001). *History: The agile manifesto.* Retrieved from http://agilemanifesto.org/

4   Beck, K., Beedle, M., van Bennekum, A., et al. (2001). *The Agile Manifesto.* Retrieved from http:// agilemanifesto.org/

5   Ibid.

6   Bjork, A. (2017). *What is agile?* Retrieved from https://docs.microsoft.com/en-us/azure/devops/learn/ agile/what-is-agile

7   Rasmusson, J. (2018). *Agile in a nutshell – Three simple truths.* Retrieved from http://www.agilenutshell. com/three_simple_truths

8   Rehkopf, M. (2018). *Atlassian agile coach: User stories.* Retrieved from https://www.atlassian.com/agile/ project-management/user-stories

9   Mahale, B. (2011). *Teamwork in agile.* Retrieved from https://www.scrumalliance.org/community/ articles/201

10   Ockerman, S. (2016). *Getting to done: Improving team collaboration.* Retrieved from https://www.scrum. org/resources/blog/getting-done-improving-team-collaboration

11   Rasmusson, J. (2018). *Agile in a nutshell – Three simple truths.* Retrieved from http://www.agilenutshell. com/three_simple_truths

12   Bjork, A. (2017). *What is agile?* Retrieved from https://docs.microsoft.com/en-us/azure/devops/learn/ agile/what-is-agile

13   Gill, S. (2013). *Key elements of a learning culture.* Retrieved from http://stephenjgill.typepad.com/ performance_improvement_b/2013/10/key-elements-of-a-learning-culture.html

14   Linders, B. (2014). *Nurturing a culture for continuous learning.* Retrieved from https://www.infoq.com/ news/2014/07/nurture-culture-learning

15   Plummer, J. (2013). *Creating a culture of learning and innovation.* Retrieved from https://www.infoq. com/articles/culture-learning-innovation

16   Rasmusson, J. (2018). *Agile in a nutshell – Three simple truths.* Retrieved from http://www.agilenutshell. com/three_simple_truths

17   Perry, M. (2018). *Only 53 US companies have been on the Fortune 500 since 1955, thanks to the creative destruction that fuels economic prosperity.* Retrieved from https://fee.org/articles/only-53-us-companies-have-been-on-the-fortune-500-since-1955-thanks-to-the-creative-destruction-that-fuels-economic-prosperity/

18   Gardasevic, S. (2015). *Fascinating new facts about Fortune 500 companies.* Retrieved from https:// domain.me/infographic-fortune-500-companies/

19   Highsmith, J. (2001). *History: The agile manifesto.* Retrieved from http://agilemanifesto.org/

20   Sidky, A. (2014). *Agile mind-set.* Retrieved from https://aspetraining.com/sites/default/files/inline-images/Ahmed%20Sidky%20Agile%20Mindset.png

21   Sidky, A. (2014). *Agile mind-set.* Retrieved from https://aspetraining.com/sites/default/files/inline-images/Ahmed%20Sidky%20Agile%20Mindset.png

22   Gardasevic, S. (2015). *Fascinating new facts about Fortune 500 companies*. Retrieved from https://domain.me/infographic-fortune-500-companies/

23   Siemens. (2017). *African digitalization maturity report 2017*. Retrieved from http://www.siemens.co.za/pool/about_us/Digitalization_Maturity_Report_2017.pdf

24   Ibid.

25   Scaled Agile (2018). *SAFe case study: Standard Bank*. Retrieved from https://www.scaledagileframework.com/standard-bank-case-study/

Chapter 2 Endnotes

26   Crous, W. (2018). The *Move towards agile and agile HR*. Retrieved from http://www.kr.co.za/_blog/knowledge-resources/post/the-move-towards-agile-and-agile-hr/

27   Rouse, M. (2016). *Contingent workforce*. Retrieved from https://searchcio.techtarget.com/definition/contingent-workforce

28   OCG Consulting. (2016). *The rise of the contingent workforce*. Retrieved from https://www.prominence.social/wp-content/uploads/2017/01/OCG-Contingent-Whitepaper-No-Bleed.pdf

29   Upwork and Freelancers Union. (2017). *Freelancing in America: 2017*. Retrieved from https://www.upwork.com/press/2017/10/17/freelancing-in-america-2017/

30   Dooley, C. (2017). *History of contingent labor: Welcoming a new, elastic workforce*. Retrieved from http://www.peopleticker.com/news/history-of-contingent-labor-welcoming-a-new-elastic-workforce

31   Goodwin, T. (2015). *The Battle Is For The Customer Interface*. Retrieved from https://techcrunch.com/2015/03/03/in-the-age-of-disintermediation-the-battle-is-all-for-the-customer-interface/

32   Manyika, J., Lund, S., Robinson, K., Valentino, J. & Dobbs, R. (2015). *Connecting talent with opportunity in the digital age*. Retrieved from https://www.mckinsey.com/featured-insights/employment-and-growth/connecting-talent-with-opportunity-in-the-digital-age

33   Pofeldt, E. (2015). *Elance-oDesk becomes 'upwork' In push to build $10B In freelancer revenues*. Retrieved from https://www.forbes.com/sites/elainepofeldt/2015/05/05/elance-odesk-becomes-upwork-today-odesk-brand-gets-phased-out/#20c69d3a51f5

34   Ibid.

35   Freelancer. (2018). *Company overview*. Retrieved from https://www.freelancer.co.za/about

36   Wikipedia. (2018, December 10). *Freelancer.com*. Retrieved from https://en.wikipedia.org/wiki/Freelancer.com

37   Wikipedia. (2017, November 17). *Fiverr*. Retrieved from https://en.wikipedia.org/wiki/Fiverr#cite_note-Fiverr_TNW-4

38   G2 Crowd. (2018). *Freelance-platforms*. Retrieved from https://www.g2crowd.com/categories/freelance-platforms

39   Ibid.

40   Ibid.

41   Rampton, J. (2017). *Employers Are Paying Freelancers Big Bucks for These 25 In-Demand Skills*. Retrieved from https://www.entrepreneur.com/article/294718

42   Reber, A.S. (1985). *The Penguin dictionary of psychology*. London & New York: Penguin Books

43   Muldowney, S. (2018). *The rise of the contingent workforce – attracting, managing and engaging transient staff*. Retrieved from https://insightsresources.seek.com.au/rise-contingent-workforce-attracting-managing-engaging-transient-staff

44   Ibid.

45   OCG Consulting. (2016). *The rise of the contingent workforce*. Retrieved from https://www.prominence.social/wp-content/uploads/2017/01/OCG-Contingent-Whitepaper-No-Bleed.pdf

46   Sands, R. & Dr. Umnub, K. (2015). *The contingent workforce: Are you aware of the traps to avoid?* Retrieved from https://www.ey.com/Publication/vwLUAssets/ey-the-contingent-workforce/%24File/ey-the-contingent-workforce.pdf

47   Amidon, S. (2017). *The Biggest Risks in Your Contingent Workforce Program (and How to Mitigate Them).* Retrieved from https://www.beeline.com/blog/biggest-risks-contingent-workforce-program-mitigate/

48   OCG Consulting. (2016). *The Rise of the Contingent Workforce.* Retrieved from https://prominence.social/wp-content/uploads/2017/01/OCG-Contingent-Whitepaper-No-Bleed.pdf

Chapter 3 Endnotes

49   Bussin, M. (2014). *Remuneration and Talent Management: Strategic Compensation Approaches for Attracting, Retaining and Engaging Talent.* Bryanston: Knowres Publishing, p.126.

50   Amundson, N.E. (2007). The influence of workplace attraction on recruitment and retention. *Journal of Employment Counseling,* 44(4), pp.154-162.

51   Business Dictionary. (2018). *Employee retention.* Retrieved from http://www.businessdictionary.com/definition/employee-retention.html

52   Bartlett, C. & Ghoshal, S. (2002). "Building competitive advantage through people". *MIT Sloan Management Review, 43*(2), pp.34-41.)

53   Bridger, E. (2018). *Employee engagement: A practical introduction.* London: Kogan Page.

54   Chua, A. (2018). *Are you working for a purpose-driven organisation?* Retrieved from https://leaderonomics.com/business/purpose-driven-organisation

55   Gallup. (2017). *State of the global workplace.* Retrieved from http://www.gallup.com/services/178517/state-global-workplace.aspx

56   Franz, A. (2017). *Employee engagement – a confluence of passion and purpose.* Retrieved from https://www.cx-journey.com/2017/07/employee-engagement-confluence-of.html

57   Ibid.

58   Sinek, S. (2009). *Start with why: how great leaders inspire action.* New York: Portfolio.

59   Ibid.

60   Ibid.

61   Thiran, R. (2013). *"Why how what" inside-out leadership: The difference between winners and losers.* Retrieved from https://leaderonomics.com/leadership/why-how-what-inside-out-leadership-the-difference-between-winners-and-losers

62   Pontefract, D. (2017). *The key differences between culture, purpose and employee engagement.* Retrieved from https://www.forbes.com/sites/danpontefract/2017/11/04/the-key-differences-between-culture-purpose-and-employee-engagement/#8b1e6993cbdb

63   Thiran, R. (2013). *"Why how what" inside-out leadership: The difference between winners and losers.* Retrieved from https://leaderonomics.com/leadership/why-how-what-inside-out-leadership-the-difference-between-winners-and-losers

64   Hakimi, S. (2015). *Why purpose-driven organisations are often more successful.* Retrieved from https://www.fastcompany.com/3048197/why-purpose-driven-companies-are-often-more-successful

65   Vaccaro, A. (2014). *How a sense of purpose boosts engagement.* Retrieved from https://www.inc.com/adam-vaccaro/purpose-employee-engagement.html

66   Biro, M.M. (2016). *Purpose or engagement? Is One Better Than the Other?* Retrieved from https://talentculture.com/purpose-or-engagement-is-one-better-than-the-other/

67   Heskett, J., Sasser, WE. Jr., & Schlesinger, L. (1997). *The Service Profit Chain: How Leading Companies Link Profit and Growth to Loyalty, Satisfaction, and Value.* New York: Free Press.

68   Pontefract, D. (2017). *The key differences between culture, purpose and employee engagement*. Retrieved from https://www.forbes.com/sites/danpontefract/2017/11/04/the-key-differences-between-culture-purpose-and-employee-engagement/#8b1e6993cbdb

69   Stern, S. (2011). *The importance of creating and keeping a customer*. Retrieved from https://www.ft.com/content/88803a36-f108-11e0-b56f-00144feab49a

70   Hurst, A. (2016). *The purpose economy*. Boise, Idaho: Elevate, USA.

71   Google. (n.d.). *About Google*. Retrieved from https://www.google.com/about/

72   Bock, L. (2011). *Passion, Not Perks*. Retrieved from: https://www.thinkwithgoogle.com/marketing-resources/passion-not-perks/

73   Apple Inc. (n.d.). *About Apple*. Retrieved from https://www.apple.com/

74   Chowdhry, A. (2013). *Lessons Learned From 4 Steve Jobs Quotes*. Retrieved from https://www.forbes.com/sites/amitchowdhry/2013/10/05/lessons-learned-from-4-steve-jobs-quotes/#413316184f69

75   21stCentury. (n.d.). *About 21st Century*. Retrieved from https://www.21century.co.za/

76   Unilever. (n.d.). *The Unilever Sustainable Living Plan*. Retrieved from https://www.unilever.co.za/sustainable-living/the-unilever-sustainable-living-plan/

77   The Walt Disney Company. (n.d.). *About the Walt Disney Company*. Retrieved from https://www.thewaltdisneycompany.com/about/

78   Nike. (n.d.). *Our Mission*. Retrieved from https://about.nike.com/

79   EY. (n.d.). *Who we are*. Retrieved from https://www.ey.com/en_gl/who-we-are

80   Levine, S.R. (2017). *Employee development for engagement and purpose*. Retrieved from https://eduleader.com/2017/07/22/employee-development-for-engagement-and-purpose/

81   Goodreads. (n.d.). *Quotable quote*. Retrieved from https://www.goodreads.com/quotes/384067-if-you-want-to-build-a-ship-don-t-drum-up

82   Pontefract, D. (2017). *The key differences between culture, purpose and employee engagement*. Retrieved from https://www.forbes.com/sites/danpontefract/2017/11/04/the-key-differences-between-culture-purpose-and-employee-engagement/#8b1e6993cbdb

Chapter 4 Endnotes

83   Veldsman, T. *(2018). People Professional of tomorrow, Part I.* Retrieved from *www.linkedin.com.*

84   Morgan, G. (2006). *Images of organization*. London & New Delhi: Sage.

85   Ibid., pp.40-41.

86   Burns, T. & Stalker, G.M. (1961). *The Management of Innovation. London:* Tavistock.

87   Morgan, G. (2006). *Images of organization*. London & New Delhi: Sage, pp.44-45.

88   McGregor, D. (1960). *The human side of enterprise. New York: McGraw-Hill.*

89   Thoren, P. (2017). *Agile people. Texas: Lionscrest.*

90   Morgan, G. (1986). *Images of Organization*. London: Sage.

91   Taylor, F.W. (1911). *Principles of scientific management. New York: Harper and Brothers Publishers.*

92   McKinsey. (2018). *The 5 trademarks of agile organizations*. Retrieved from https://www.mckinsey.com/business-functions/organization/our-insights/the-five-trademarks-of-agile-organizations

93   Ibid.

94   Ibid.

95   Thoren, P. (2017). *Agile people. Texas: Lionscrest.*

96   Rigby, D. K., Sutherland, J., & Hirotaka, T. (2016). Embracing agile: How to master the process that's transforming management. *Harvard Business Review, 94*(5), pp.40-50.

97   Ulrich, D. (2018). *Agility: The new response to dynamic change.* Retrieved from www.linkedin.com

98   De Smet, A. & Gagnon, C. (2018). *Organizing for the age of urgency.* Retrieved from https://www. mckinsey.com/business-functions/organization/our-insights/organizing-for-the-age-of-urgency

99   Veldsman, T. *(2018). People Professional of tomorrow, Part I.* Retrieved from www.linkedin.com.

100   McKinsey. (2018). *The 5 trademarks of agile organizations.* Retrieved from https://www.mckinsey.com/ business-functions/organization/our-insights/the-five-trademarks-of-agile-organizations

101   Wheatley, M. J. (2006). *Leadership and the new science: Discovering order in a chaotic world (3rd ed.). San Francisco: Berrett-KoehlerPublishers.*

102   SEMCO Style Institute. (n.d.). *Make Work Awesome.* Retrieved from https://semcostyle.org/.

103   Semler, R. (1993). *Maverick. New York: Warner books. London: Century.*

104   Semler, R. (2003). *The Seven Day Weekend. London: Random House.*

Chapter 5 Endnotes

105   Colvin, G. (2001). *Fortune – The New Future.* Retrieved from: http://www.semco.com.br/en/download/ media-coverage/fortune-november-2001.pdf

106   Borges, I. (2018). How Semco Introduced it's Participatory Culture. Retrieved from: https://semcostyle. org/articles/2018/09/how-semco-introduced-it-s-participatory-culture

107   Conboy, K., Coyle, S., Wang, X. & Pikkarainen, M. (2011). *People over process: Key people challenges in agile development.* Unpublished article. Retrieved from https://ulir.ul.ie/bitstream/ handle/10344/639/2010-conboy-people.pdf?sequence=2.

108   Chartered Institute of Personnel and Development (CIPD). (2018). *Fact sheet: Organisational Development.* Retrieved from https://www.cipd.co.uk/knowledge/strategy/organisational-development/factsheet.

109   Huntington, J., Gillam, S. & Rosen, R. (2000). Organisational development for clinical governance. *British Medical Journal, 321*, p.679.

110   Fox, H.L. (2013). The promise of organisational development in non-profit human services organisations. *Organisational Development Journal, 31*(2), pp.1-100.

111   French, W. & Bell, C. (1999). *Organisation development: Behavioural science Interventions for organisation improvement.* Portland: Book News, Inc.

112   Emond, L. (2018). *Agility Is Both Structural and Cultural at Roche.* Retrieved from https://www.gallup. com/workplace/243167/agility-structural-cultural-roche.aspx?utm_source=workplacenewsletter&utm_ medium=email&utm_campaign=WorkplaceNewsletter_October_102318&utm_ content=getinsiderinsights-CTA-2&elqTrackId=3C1D0013998BC8D1290AAE292CFD4BA9&elq=25580c e50f8f46289d7b3b9c046ef6eb&elqaid=312&elqat=1&elqCampaignId=80.

113   Iles, P. & Yolles, M. (2003). Complexity, HRD and Organisation Development: Towards a Viable Systems Approach to Learning, Development and Change. In Lee, M. (Ed.). *HRD in A Complex World, Studies in Human Resource Development.* Oxford: Routledge.

114   Burke, W. W. (1994). *Organisation development: A process of learning and changing.* Reading, MA: Addison-Wesley.

115   Mumford, A. (1991). Individual and organisational learning: the pursuit of change. *Industrial and Commercial Training, 23*(6), pp.24-31.

116   Grieves, T. & Redman, T. (1999). Living in the shadow of OD: HRD and the search for identity, *Human Resource Development International, 2*(2), pp.81-102.

117   McLean, G.N. (2005). *Organization development: Principles, processes, performance.* San Francisco, CA: Berrett-Koehler Publishers.

118 Ramlall, S. (2004). A review of employee motivation theories and their implications for employee retention within organizations. *The Journal of American Academy of Business. 5*(1/2) 52-63.

119 McLaughlin, J. (2018). *What is organizational culture? Definition & characteristics.* Retrieved from https://study.com/academy/lesson/what-is-organizational-culture-definition-characteristics.html

120 McAvoy, J. & Butler, T. (2009). The role of project management in ineffective decision making within agile software development projects. *European Journal of Information Systems, 18,* pp.372-383.

121 Gallup. (2018). *The real future of work: The agility issue.* Retrieved from https://www.gallup.com/workplace/241295/future-work-agility-download.aspx?g_source=link_WWWV9&g_medium=related_tile1&g_campaign=item_243167&g_content=The%2520Real%2520Future%2520of%2520Work%3a%2520Agility%2520Issue.

122 Iles, P. & Yolles, M. (2003). Complexity, HRD and Organisation Development: Towards a Viable Systems Approach to Learning, Development and Change. In Lee, M. (Ed.). *HRD in A Complex World, Studies in Human Resource Development.* London: Routledge.

123 Chandler, R. G. (2018). Agile teams create agile learning organisations. *Chief Learning Officer,* July/August, p.56.

124 Pantouvakis, A. & Bouranta, N. (2017). Agility, organisational learning culture and relationship quality in the port sector. *Total Quality Management, 28*(4), pp.366-378.

125 Ibid.

126 Chandler, R. G. (2018). Agile teams create agile learning organisations. *Chief Learning Officer, July/August, pp.55-65.*

127 Ibid.

128 Gallup. (2018). *The real future of work: The agility issue.* Retrieved from https://www.gallup.com/workplace/241295/future-work-agility-download.aspx?g_source=link_WWWV9&g_medium=related_tile1&g_campaign=item_243167&g_content=The%2520Real%2520Future%2520of%2520Work%3a%2520Agility%2520Issue.

129 Ibid.

130 Omar, M., Khasasi, N.L.A., Abdullah, S.L.S., Hashim, N.L., Romli, R. & Katuk, N. (2018). Defining skill sets requirements for agile scrum team formation. *Journal of Engineering and Applied Sciences, 13,* pp.784-789. DOI: 10.3923/jeasci.2018.784.789.

131 Ibid.

132 Ibid.

133 Conboy, K., Coyle, S., Wang, X. & Pikkarainen, M. (2011). *People over process: Key people challenges in agile development.* Retrieved from https://ulir.ul.ie/bitstream/handle/10344/639/2010-conboy-people.pdf?sequence=2.

134 Ibid.

Chapter 6 Endnotes

135 Bussin, M. & Smit, E. (2015). *Performance management revealed.* Retrieved from http://www.sara.co.za/sara/file%20storage/Documents/articles/2015%20March%20-%20Performance%20Management%20Revealed%20-%20Dr%20M%20Bussin%20and%20Elmien%20Smit.pdf

136 Armstrong, M. (1996). *A handbook of personnel management practice.* London: Kogan.

137 WorldatWork. (2018). *Performance management.* Retrieved from https://www.worldatwork.org/

138 CIPD. (2016). *Could do better? Assessing what works in performance management.* Retrieved from https://www.cipd.co.uk/knowledge/fundamentals/people/performance/what-works-in-performance-management-report

139 Bersin, J. (2014). *The Top 10 disruptions in HR technology: Ignore them at your peril.* Retrieved from https://www.forbes.com/sites/joshbersin/2014/10/15/the-top-ten-disruptions-in-hr-technology-ignore-them-at-your-peril/#6781f6afa1fa

140 Krebsbach, H. (2017). *Building a winning team with agile performance management.* Retrieved from https://www.atlassian.com/blog/agile/building-winning-team-agile-performance-management

141 Rehkopf, M. (2018). *What is a Kanban board?* Retrieved from https://www.atlassian.com/agile/kanban/boards

142 Techopedia. (2018). *What is Scrum Sprint?* Retrieved from https://www.techopedia.com/definition/13687/scrum-sprint

143 Erikson Coaching International. (2018). *What is the difference between coaching and managing?* Retrieved from https://erikson.edu/coaching-and-managing-differences

144 Gray, S., Stewart, A., Anderson, B., Handley, C., Bray, D., Darling, P. & Chivers, W. (n.d.). *Best practice guidelines in 360 degree feedback.* Roehampton: University of Surrey.

145 Engagedly. (2018). *360 degree feedback overview.* Retrieved from https://engagedly.com/360-degree-performance-review-multirater/

146 Radigan, D. (2018). *Stand-ups for agile teams.* Retrieved from https://www.atlassian.com/agile/scrum/standups

147 Agile Connection. (2015). *The scrum daily standup meeting: Your questions answered.* Retrieved from https://www.agileconnection.com/article/scrum-daily-standup-meeting-your-questions-answered

148 Adobe. (2018). *The story of check-in.* Retrieved from https://www.adobe.com/check-in.html

149 Burkus, D. (2017). *How Adobe structures feedback conversations.* Retrieved from https://hbr.org/2017/07/how-adobe-structures-feedback-conversations?utm_source=twitter&utm_medium=social&utm_campaign=hbr

150 Hearn, S. (2016). *Performance management case studies: Revolutionaries and trail blazers.* Retrieved from https://clearreview.com/top-5-performance-management-case-studies/

151 Powell, J. & Chapman, B. (2014). *Top pitfalls of agile development.* Retrieved from https://www.credera.com/blog/business-insights/top-pitfalls-agile-development/

152 Ibid.

Chapter 7 Endnotes

153 Gallo, A. (2013). *How to Reward Your Stellar Team.* Retrieved from https://hbr.org/2013/08/how-to-reward-your-stellar-tea

154 Cohn, M. (2016). *How to reward agile teams.* Retrieved from https://www.mountaingoatsoftware.com/blog/how-to-reward-agile-teams

155 Goldston, N.J. (2017). *Five Creative Ways to Reward your Team now and All Year Long.* Retrieved from https://www.forbes.com/sites/njgoldston/2017/12/11/five-creative-ways-to-reward-your-team-now-and-all-year-long/#3f94abc77a77

156 Eyholzer, F. (2016). *Rethinking reward and recognition (agile HR – Q&A series part 3).* Retrieved from https://www.scrumalliance.org/agilecareers/careersblog/december-2016/rethinking-reward-and-recognition-(agile-hr-%E2%80%93-q-a

157 Friederichs, E. (2008). *Fairness and Transparency Are Essential.* Retrieved from https://www.di.net/articles/fairness_transparency_are_essential/

158 Scaled Agile. (2017). *Principle #8 – Unlock the intrinsic motivation of knowledge workers.* Retrieved from https://www.scaledagileframework.com/unlock-the-intrinsic-motivation-of-knowledge-workers/

159 Ibid.

160 Ibid.

161 Davis, W.W. (2017). *Annual performance reviews and bonuses are anti-agile. Let's abolish them.* Retrieved from https://www.linkedin.com/pulse/annual-performance-reviews-bonuses-anti-agile-lets-abolish-davis

162 Ibid.

163 Scaled Agile. (2017). Principle #8 – Unlock the intrinsic motivation of knowledge workers. Retrieved from https://www.scaledagileframework.com/unlock-the-intrinsic-motivation-of-knowledge-workers/

164 Cappelli, P. & Tavis, A. (2018). *HR goes agile.* Retrieved from https://hbr.org/2018/03/the-new-rules-of-talent-management#hr-goes-agile

165 Ibid.

Chapter 8 Endnotes

166 Farson, R. (2008). *The Power of Design: A Force for Transforming Everything.* Norcross, GA: Greenway Communications.

167 Beaujean, M., Davidson, J. & Madge, S. (2006). *The 'moment of truth' in customer service.* Retrieved from https://www.mckinsey.com/business-functions/organization/our-insights/the-moment-of-truth-in-customer-service

168 Design Kit. (n.d.). *Methods.* Retrieved from http://www.designkit.org/methods

169 Bersin, J. (2016). *Predictions for 2017: The Digital World of Work is Transforming Management, Human Resources and People Practices.* Retrieved from https://www.prnewswire.com/news-releases/predictions-for-2017-the-digital-world-of-work-is-transforming-management-human-resources-and-people-practices-300374432.html

170 Meister, J. (2016). *Cisco HR Breakathon: Reimagining the Employee Experience.* Retrieved from https://www.forbes.com/sites/jeannemeister/2016/03/10/the-cisco-hr-breakathon/#6dac4e07f5ee

171 Naiman, L. (2017). *Why Your HR Department Should Embrace Design Thinking.* Retrieved from https://www.inc.com/linda-naiman/6-ways-hr-applies-design-thinking-to-deliver-engaging-employee-experiences.html

172 Meister, J. (2015). *The Future of Work: Airbnb CHRO Becomes Chief Employee Experience Officer.* Retrieved from https://www.forbes.com/sites/jeannemeister/2015/07/21/the-future-of-work-airbnb-chro-becomes-chief-employee-experinece-officer/#35ce0ca54232

173 Ibid.

174 Coy, C. (2017). *A Day in the Life of an Employee Experience Manager and Specialist.* Retrieved from https://www.cornerstoneondemand.com/rework/day-life-employee-experience-manager-and-specialist

175 Ibid.

176 Kolawole, E. (2016). *From empathy to community.* Retrieved from *http://www.niemanlab.org/2016/12/from-empathy-to-community*

177 Lee, H. (1960). *To Kill a Mockingbird.* Philadelphia: J.B. Lippincott & Co.

178 Ries, E. (2011). *The Lean Startup: How Today's Entrepreneurs use Continuous Innovation to Create Radically Successful Businesses.* New York: Crown Publishing Group

179 Ibid.

180 Hogle, P. (2017). *MVP Is the Key to Agile Project Management.* Retrieved from https://www.learningsolutionsmag.com/articles/2308/mvp-is-the-key-to-agile-project-management

Chapter 9 Endnotes

181 Genseler. (2016). *Gensler releases U.K. workplace survey 2016 findings.* Retrieved from https://www.gensler.com/news/press-releases/uk-workplace-survey-2016-findings

182 Ibid.

183 Home Designs Insights. (2018). *Office cubicle walls.* Retrieved from http://www.hotelresicolibri.com/cubicle-walls-storage-accessories-ideas/office-cubicle-walls/

184 Office Changes. (2018). *Would your business benefit from an open plan office design?* Retrieved from https://www.officechanges.com/open-plan-office-design/

185 Gensler. (2016). *U.S. Workplace Survey 2016.* Retrieved from https://www.gensler.com/news/press-releases/us-workplace-survey-2016-findings

186 McLaurin, J.P. (2016). *How can the workplace impact innovation?* Retrieved from http://www.gensleron.com/work/2016/7/8/how-can-the-workplace-impact-innovation.html

187 Aon Hewitt. (2017). *2017 Trends in Global Employee Engagement.* Retrieved from http://www.aon.com/unitedkingdom/attachments/trp/2017-Trends-in-Global-Employee-Engagement.pdf

188 Moutrey, G. (2014). *Power of Place: The Office Renaissance.* Retrieved from https://www.steelcase.com/research/360-magazine/the-privacy-crisis-issue-68/

189 Ibid.

190 Ibid.

191 Zero Gravity Tables. (2018). *Standing desks sit to stand workstation table.* Retrieved from https://www.zerogravitytables.com/standing-desks-sit-to-stand-workstation-table/

192 Moutrey, G. (2014). *Power of Place: The Office Renaissance.* Retrieved from https://www.steelcase.com/research/360-magazine/the-privacy-crisis-issue-68/

193 Genseler. (2016). *Gensler releases U.K. workplace survey 2016 findings.* Retrieved from https://www.gensler.com/news/press-releases/uk-workplace-survey-2016-findings

194 Malburg, M. (2016). *Designing spaces that work for a multigenerational workforce.* Retrieved from https://www.progressiveae.com/creating-multigenerational-spaces/

195 Moutrey, G. (2014). *Power of Place: The Office Renaissance.* Retrieved from https://www.steelcase.com/research/360-magazine/the-privacy-crisis-issue-68/

196 Turnstone. (2017). *How to create quiet spaces in the workplace.* Retrieved from https://myturnstone.com/blog/how-to-create-quiet-spaces-in-workplace/

197 Morgan, J. (2015). *How The Physical Workspace Impacts The Employee Experience.* Retrieved from https://www.forbes.com/sites/jacobmorgan/2015/12/03/how-the-physical-workspace-impacts-the-employee-experience/#4e983df1779e

198 Moutrey, G. (2014). *Power of Place: The Office Renaissance.* Retrieved from https://www.steelcase.com/research/360-magazine/the-privacy-crisis-issue-68/

199 Morgan, J. (2015). *How The Physical Workspace Impacts The Employee Experience.* Retrieved from https://www.forbes.com/sites/jacobmorgan/2015/12/03/how-the-physical-workspace-impacts-the-employee-experience/#4e983df1779e

200 Ibid.

201 Cision. (2015). Global *study connects levels of employee productivity and well being to office design.* Retrieved from https://www.prnewswire.com/news-releases/global-study-connects-levels-of-employee-productivity-and-well-being-to-office-design-300058034.html

202 Nieuwenhuis, M., Knight, C., Postmes, T. & Haslam, S. A. (2014). The relative benefits of green versus lean office space: Three field experiments. *Journal of Experimental Psychology: Applied, 20*(3), pp.199-214. http://dx.doi.org/10.1037/xap0000024

203 Cigniti Technologies. (2018). *5 benefits of employing collocated teams for agile software testing.* Retrieved from https://www.cigniti.com/blog/benefits-of-colocated-teams-for-agile-software-testing/

204 Ibid.

205 Steelcase. (2018). *Transforming IT at Steelcase: An agile case study.* Retrieved from https://www.steelcase.com/research/articles/topics/agile/agile-case-study/

Chapter 10 Endnotes

206 Turing, A.M. (1950). Computing machinery and intelligence. *Mind, 49*, pp.433-460.

207 Dalakov, G. (2018). *History of Computers.* Retrieved from https://history-computer.com/index.html

208 Boettcher, S. (1999). *Case Study: the IBM/Electric Minds 'Kasparov v. Deep Blue Chess Match'.* Retrieved from http://www.fullcirc.com/community/figalloibm.htm

209 IntelliPaat. (2017). *The unstoppable power of deep learning – AlphaGo vs. Lee Sedol case study.* Retrieved from https://intellipaat.com/blog/power-of-deep-learning-alphago-vs-lee-sedol-case-study/

210 Burgess, A. (2017). *RPA & AI.* Retrieved from https://disruptionhub.com/robotic-process-automation-artificial-intelligence/

211 Ibid.

212 Craig, A., Lacity, M. & Willocks, L. (2015). *Robotic process automation at telefónica O2.* The Outsourcing Unit Working Paper Series. Paper 15/02.

213 Fagella, D. (2018). *5 Business intelligence analytics case studies across industry.* Retrieved from https://www.techemergence.com/5-business-intelligence-analytics-case-studies-across-industry/.

214 Lunden, I. (2016). *IBM's Watson now powers AI for under armour, Softbank's Pepper robot and more.* Retrieved from https://techcrunch.com/2016/01/06/ibms-watson-now-powers-ai-for-under-armour-softbanks-pepper-robot-and-more/.

215 Welch, C. (2018). *Google just gave a stunning demo of Assistant making an actual phone call.* Retrieved from https://www.theverge.com/2018/5/8/17332070/google-assistant-makes-phone-call-demo-duplex-io-2018.

216 Brown, R. (2017). *Where is AI Headed in 2018?* Retrieved from https://blogs.nvidia.com/blog/2017/12/03/ai-headed-2018/.

217 Oxfam International. (2017). *Just 8 men own the same as half the world.* Retrieved from https://www.oxfam.org/en/pressroom/pressreleases/2017-01-16/just-8-men-own-same-wealth-half-world.

Chapter 11 Endnotes

218 Ulrich, D. (1997). *Human Resource Champions: The next agenda for adding value and delivery results.* Boston, MA: Harvard Business School Press.

219 Investopedia. (n.d.). *Human Resources (HR).* Retrieved from https://www.investopedia.com/terms/h/humanresources.asp

220 Whatishumanresources.com. (n.d.) *The historical background of human resource management.* Retrieved from http://www.whatishumanresource.com/the-historical-background-of-human-resource-management

221 Ibid.

222 Cappelli, P. &Tavis, A. (2018). *HR Goes Agile.* Retrieved from https://hbr.org/2018/03/the-new-rules-of-talent-management#hr-goes-agile

223 Shingal, T. (2018). *Managing HR roles: David Ulrich's model.* Retrieved from https://blog.mettl.com/talent-hub/managing-hr-roles-david-ulrich-model

224 Denning, S. (2011). *How strategic HR wins the keys to the C-suite.* Retrieved from https://www.forbes. com/sites/stevedenning/2011/04/06/how-strategic-hr-wins-the-keys-to-the-c-suite/#7e7258fb46fe

225 Sloan, N., Agarwal, D., Garr, S. & Pastakia, K. (2017). *Performance management: Playing a winning hand – Deloitte 2017 global human capital trends.* Retrieved from https://www2.deloitte.com/insights/us/en/ focus/human-capital-trends/2017/redesigning-performance-management.html

226 Benado, Y. (2015). *Need top talent fast? Try agile recruiting.* Retrieved from https://techbeacon.com/ need-top-talent-fast-try-agile-recruiting

227 SolutionsIQ. (2018). *Resources for agile humans.* Retrieved from https://www.solutionsiq.com/resource/ white-paper/resources-for-agile-humans/

228 Cappelli, P. &Tavis, A. (2018). *HR Goes Agile.* Retrieved from https://hbr.org/2018/03/the-new-rules-of-talent-management#hr-goes-agile

229 Aghina, W., De Smet, A., Lackey, G., Lurie, M. & Murarka, M. (2018). *The five trademarks of agile organizations.* Retrieved from https://www.mckinsey.com/business-functions/organization/our-insights/the-five-trademarks-of-agile-organizations

230 Forcano, R. (2018). *HR goes agile: A case study in BBVA.* Retrieved from https://www.bbva.com/en/ opinion/hr-goes-agile-case-study-bbva/

231 Huss, S. (2017). *The new role of HR in the agile organization.* Retrieved from http://tracks.roojoom. com/r/88327#/trek?page=2

232 Towers Watson. (2014). *Perspectives. Employee voice. Releasing voice for sustainable business success.* Retrieved from https://docplayer.net/19382130-Perspectives-employee-voice-releasing-voice-for-sustainable-business-success.html

233 Mahajan, A. (2013). *The importance of HR in agile adoption.* Retrieved from https://www.scrumalliance. org/community/articles/2013/january/the-importance-of-hr-in-agile-adoption

234 Huss, S. (2017). *The new role of HR in the agile organization.* Retrieved from http://tracks.roojoom. com/r/88327#/trek?page=2

235 Mahajan, A. (2013). *The importance of HR in agile adoption.* Retrieved from https://www.scrumalliance. org/community/articles/2013/january/the-importance-of-hr-in-agile-adoption

INDEX

www.ingramcontent.com/pod-product-compliance
Lightning Source LLC
Chambersburg PA
CBHW081504200326
41518CB00015B/2375